PATTERN-BASED COMPRESSION OF MULTI-BAND IMAGE DATA FOR LANDSCAPE ANALYSIS

Environmental and Ecological Statistics

Series Editor
G. P. Patil

Center for Statistical Ecology and Environmental Statistics
Department of Statistics
The Pennsylvania State University
University Park, PA, 16802, USA
Email address: gpp@stat.psu.edu

The Springer series **Environmental and Ecological Statistics** is devoted to the cross-disciplinary subject area of environmental and ecological statistics discussing important topics and themes in statistical ecology, environmental statistics, and relevant risk analysis. Emphasis is focused on applied mathematical statistics, statistical methodology, data interpretation and improvement for future use, with a view to advance statistics for environment, ecology, and environmental health, and to advance environmental theory and practice using valid statistics.

Each volume in the **Environmental and Ecological Statistics** series is based on the appropriateness of the statistical methodology to the particular environmental and ecological problem area, within the context of contemporary environmental issues and the associated statistical tools, concepts, and methods.

Additional information about this series can be obtained from our website:
www.springer.com

PATTERN-BASED COMPRESSION OF MULTI-BAND IMAGE DATA FOR LANDSCAPE ANALYSIS

by

Wayne L. Myers
School of Forest Resources
The Pennsylvania State University, USA

and

Ganapati P. Patil
Center for Statistical Ecology and Environmental Statistics
The Pennsylvania State University, USA

 Springer

ISBN-13: 978-1-4419-4271-5   e-ISBN-13: 978-0-387-44439-0

Printed on acid-free paper.

9 8 7 6 5 4 3 2 1

springer.com

Dedication

This monograph is dedicated to Faith and Lalit, not as a work but as a labor of love, in the same spirit with which our wives have so lovingly supported us over the years during which these analytical approaches and their predecessors were in formative and convergent stages. They have patiently encouraged us, and have often assumed additional tasks while we were occupied or preoccupied with these concepts and endeavors. Likewise, they have tolerated our travels and extended absences when circumstances made it difficult for them to accompany us. They have frequently taken charge of domestic matters and sometimes serve as our surrogates with family and friends while we concentrate attention on analysis, brainstorming, or professional networking. So also for the numerous delays and rescheduling of personal projects that they have gracefully accommodated. Then they celebrate with us the auspicious outcomes of these ventures. Our appreciation of them, and our gratitude to them, go way beyond words. We salute them as best friends.

Preface

We offer here a non-conventional approach to multivariate image-structured data for which the basis is well tested but the analytical ramifications are still unfolding. Although we do not formally pursue them, there are several parallels with the nature of neural networks. We employ a systematic set of statistical heuristics for modeling multivariate image data in a quasi-perceptual manner. When the human eye perceives a scene, the elements of the scene are segregated heuristically into components according to similarity and dissimilarity, and then the relationships among the components are interpreted. Similarly, we segregate or segment the scene into hierarchically organized components that are subject to subsequent statistical analysis in many modes for interpretive purposes. We refer to the segregated scene segments as patterns, since they provide a basis for perception of pattern. Since they are also hierarchically organized, we refer to them further as polypatterns. This leads us to our acronym of Progressively Segmented Image Modeling As Poly-Patterns (PSIMAPP). Likewise, we formalize our approach in terms of pattern processes and segmentation sequences. In alignment with the terminology of image analysis, we refer to our multivariate measures as being signal bands.

There are several aspects of analytical advantage in our approach. We model multi-band image data in terms of A-level aggregated and B-level base patterns, with each level occupying an amount of computer media equivalent to one signal band. Therefore, our models are parsimonious relative to image data and can confer informational compression when there are three or more signal bands. The models allow approximate restoration of the image data, but since some residuals remain, the compression is of a lossy nature which can obviate some of the issues regarding redistribution and other proprietary provisions on the initial image information. An additional advantage is that the A-level model is completely compatible with facilities for handling raster maps in geographic information systems (GIS) that constitute the conventional contemporary computational capability for conveying spatially specific geographical information. A substantial suite of advantageous aspects are encompassed in our 4CS pattern perspective of *segmenting signal, spatial* and *sequential* (temporal) information for consideration of *contrast, content, context* and *change* at a landscape level spanning several spatial scales.

Convenient contrast control in portraying prominent patterns is provided by preparing pictorial presentations through tonal transfer tables. Depictions are done by developing columns of transfer tables using pattern properties to produce palettes for patterns instead of the slower scanning of scenes. This supports special selection of scene segments for showcasing or suppression. Pictorial presentations are managed mostly as mosaic maps of patterns. Investigation of innovative integration of several signals as image indicators can be done in spreadsheet software instead of complex computational configurations of pixel processing packages. Integrative indexing culminates in computing composite intensity as average across signal bands in each of the A-level patterns and using the rank order of that intensity as an identifying index for the particular pattern. The map of these index identifiers can be displayed directly as a single-band image of what we call *ordered overtones* for the A-level patterns of our model. Image information processing packages are publicly available for downloading without charge that can be used to portray pattern pictures.

Parsing of patterns permits auto-adaptive contrast control, which automatically adjusts contrast relations among patterns by an iterative algorithm using statistical criteria that would be unworkable for initial image information. This is in the nature of a statistically guided sequential linear 'stretch' with 'saturation'. Stretching involves proportional extension of dynamic range of signals, whereas saturation entails truncation of signal ranges. Contrast is perceived as differences in an image, with differences due to variation which is expressed statistically as variance. Increasing contrast is associated with increasing variance of intensities in the image. Truncation by saturation induces areas of uniformity in high intensity while increasing differences for lower intensities. There is also an issue of tonal balance, whereby the average intensity should have intermediate brightness. The adaptive algorithm proceeds with modification by saturation as long as variance increases and average intensity remains at or below a preset level.

Polypatterns provide for interactive interpretive classification of scene content such as land cover for thematic mapping in a manner that would not be accommodated by conventional image analysis software systems. As with pictorial presentations, this can be accomplished with software for personal computers that is publicly available for downloading without charge. Polypatterns also accommodate algorithmic approaches that are usually called *supervised* and *unsupervised* classification, as well as adaptive strategies that hybridize the two approaches. Transfer tables for pictorial presentation can also be integrated into the mapping methods so that particular patterns can be shown as categorical color imbedded directly into grayscale renderings of the surrounding scenery.

Contextual considerations concern the fragmentation, juxtaposition and interspersion of the pattern components comprising the image model. Much of the visual perception of context in a picture is geometrically generated by mental models from information that is implicit rather than explicit in the image. Crucial contextual capability comes in segregating segments to enable explicit analysis of their arrangements. With the modeled mapping of pattern segments it becomes possible to investigate their joint DIStribution of POSITION (or DISPOSITION) in multiple modes. One line of inquiry that would not be available for ordinary images is to examine *edge affinities*, or preferential juxtaposition of particular patterns relative to expectations in a RANdom landSCAPE (or RANDSCAPE). By comparing edge affinities and signal similarities as dual domains, it becomes possible to explore several generations of generalization to produce primary patterns of the landscape.

Innovative investigation of multi-scale landscape characteristics can be conducted by compiling frequency profiles of patterns in nested blocks of cellular image elements (pixels). Such profiles capture composite characteristics of vicinities in the various settings of a landscape. A newly developed measure of profile differences as *accordion distance* allows guided grouping of similar settings across landscapes, and thus systematic study of landscape linkages. Equally innovative is investigation of compositional components as families of patterns through partial ordering and rank range runs.

The topological spatial structure of the step surface formed by the model's multivariate ordered overtones can be expressed in terms of echelon hierarchies. Topological territories and transitions can thereby be isolated for intensive investigation. Restored regions of the models can be subtracted from the initial image information for purposes of multivariate detrending so that assumptions of homogeneity and isotropy are more applicable for analysis of residuals by the methods of spatial statistics.

A principal components approach can be adapted to explore the signal structure of the models in relation to that of the original image data. This entails compiling covariance and correlation matrices for restorations of the signal bands from the models. Extracting eigenvalues and eigenvectors from these matrices give what we call *principal properties* in parallel to and for comparison with principal components of the original image information.

Much of the impetus for our investigations has come in the interest of detecting differences between instances of imaging in order to capture change. The classic case is for sequences of scenes taken over time with the same sensor. In this relatively simple scenario, the signal smoothing from pattern processes helps to make perturbed patches prominent. Additional advantage and innovative improvement comes from matching mosaics of patterns in paired instances of imaging. Perturbation of patterns in matched mosaics provides difference detection even with a shift in sensing systems. Segregating sets of signals from a sensor also allows assessment of differences in detectors. Still more substantial is the second-order advantage available by polypattern processing after compositing indicators of change across multiple instances of imaging to track temporal trajectories in landscape dynamics.

We have provided a substantial sampling of pattern pictures to illustrate applications of our approaches, but have been considerably constrained in do so by need to forego color in order to control costs. Color can make an amazing difference in preparation of pattern pictures. Likewise, the simpler printing processes do not provide the richness of resolution that is available directly on a computer display screen. We ask that you do a web search for PSIMAPP to access available software for applications. The web is also a medium through which we can also offer case studies in color, although still not with the resolution capabilities of a computer console.

Contents

Contributing Authors:

Wayne L. Myers, Ph.D.
Professor of Forest Biometrics, School of Forest Resources
Director, Office for Remote Sensing and Spatial Information Resources
Penn State Institutes of Environment
The Pennsylvania State University

Ganapati P. Patil, Ph.D., D.Sc.
Distinguished Professor of Mathematical Statistics, Department of Statistics
Director, Center for Statistical Ecology and Environmental Statistics
Department of Statistics
The Pennsylvania State University

Acknowledgments

We appreciatively acknowledge an array of assistance spanning several sorts. First on this list is the late Charles Taillie whose intellectual inspiration and insight permeated our cooperative working arrangements for so many years. Genuine genius seems to be the best way of phrasing his phenomenal mathematical mastery. Close behind are a number of former graduate students, including but not limited to Dr. Glen Johnson, Joe Bishop, Dr. Ningning Kong, Dr. Shih-Tsen Liu, Dr. Brian Lee, Dr. Somboon Kiratiprayoon, Francis Beck, Apinya Ramakomud and Dr. Salleh Mohd. Nor. Emily Hill receives recognition for her capable conversion of diverse document files to camera ready final form in meeting our timelines.

Agency, organizational and institutional acknowledgments are also in order. The Penn State Institutes of Environment, and the School of Forest Resources through the Pennsylvania Agricultural Experiment Station have provided continuing support. The national GAP Analysis Program of biodiversity assessment furnished foundational support as it migrated through agency settings from the U.S. Fish and Wildlife Service to the U.S. Geological Survey. This was augmented by funding for change detection work through NASA Goddard Space Flight Center and the encouragement of Dr. Darrel Williams. The National Science Foundation and Environmental Protection Agency partnering programs in Water and Watersheds initiatives enabled emphasis on landscapes and linkages across scales. Most recently, the National Geospatial Intelligence Agency contributed to contextual concepts and computations. Finally, KUSTEM University in Malaysia was a gracious host for Fellowship work by which to make the manuscript manifest.

Also, all along, the Center for Statistical Ecology and Environmental Statistics in the Department of Statistics at Penn State has provided conducive work facilities. And at the turn of this century, the National Science Foundation Digital Government Research Program and the Environmental Agency STAR Grants Program have been supporting related research, training, and outreach initiatives.

1 Innovative Imaging, Parsing Patterns and Motivating Models

It seems that much of science has a propensity to concern itself with microscopic scales by focusing on genomic, molecular, atomic and subatomic phenomena from a perspective of reductionism. The sciences of ecology and sustainable environment, however, must counterbalance this by investigating patterns of interaction in space and time at larger scales covering landscapes and regions. This is particularly true for the ecological discipline of *landscape ecology* (Forman & Godron, 1986; Forman, 1995; McGarigal & Marks, 1995; Turner, Gardner & O'Neill, 2001; Myers et al., 2006). Image data and innovative imaging play important roles in such investigations by serving as macroscopes to reveal patterns of arrangement and change over substantial spatial extents at several scales. It is to these macroscopic scales that we turn our attention in this work, with particular emphasis on patterns in images.

The data that drive our imaging must be *synoptic*, by which we mean that it is possible and sensible to ascribe a value to any location within the geographic extent of interest – whether by direct determination or by interpolation. The data need not necessarily arise, however, from conventional remote sensing based on signal scenarios involving radiant energy of the electromagnetic spectrum. One of our thrusts is innovative imaging whereby a variety of environmental indicators are cast in the manner of multiple 'bands' of images so that spatial patterns and interactions having ecological implications can be more readily detected, analyzed, and tracked over time.

With this pattern purview, we focus on advanced approaches to parsing patterns into components so that the spatial arrangements of the components can be subjected to systematic study both statistically and structurally. Patterns are often found to have dominant and subordinate contributions that express at different scales, in what might be considered as more evident overtones and more subtle undertones. We seek to facilitate cognizance and characterization of these multi-scale manifestations of spatial relations in landscapes.

Principal parts of patterns and their area arrangements present possibilities of parsimonious packaging as multi-models of images that confer compression as well as enabling enhancements of pattern groupings for pictorial presentations in a more flexible fashion than regular renderings. If pattern positions are empirically expressed in the mode of modeling, then alternative analysis of spatial structure can be done directly by means of the model without recourse to restoration. Analytical approaches to crucial concerns of contrast, content, context and change are also augmented.

Introduction of image innovations will be facilitated by an opening overview of certain conventional concepts regarding signals, scenes and sensors. This provides a path for pursuit of Progressively Segmented Image Models As Poly-Patterns (PSIMAPP), with poly-pattern henceforth being written as just *polypattern*.

1.1 Image Introductory

Images are composed of *pixels* (picture elements) forming a regular lattice L of area units that limit the spatial resolution of the image. Lattice elements of essentially equivalent nature are called *cells* in the terminology of *geographic information systems* (GIS) for mapping landscapes (Bolstad, 2005; Chrisman, 2002; DeMers, 2000). The terms *cell* and *pixel* will be used interchangeably in the course of this presentation, although the latter term usually implies a *rectangular* lattice (*raster* arrangement) whereas the former might be non-rectangular as for example hexagonal. In the terminology of *spatial statistics*, the area comprising a lattice element is its *spatial support* (Schabenberger & Gotway, 2005). Associated with each lattice element (pixel or cell) is an ordered list (vector) of numeric values with regard to particular *properties*. We will refer to values comprising such vectors as *signals* in a generic manner. As used here, a signal *band* is a particular component of the signal vector (i.e., specific position in the list of signal values). Accordingly, we denote the set of all distinct signal vectors by V without reference to particular pixel positions.

An image is formed electronically by translating the values of one or more bands into intensities of gray-tone or color for displaying in the respective lattice positions of the pixels. Fig. 1.1 is an image formed by rendering a band of red light recorded by a sensor aboard a satellite (MSS sensor of Landsat satellite) as gray-tone intensities. An image of this sort is based on reception of a physical signal by a remotely positioned sensing device, and thus falls under the rubric of *remote sensing*. Since remote

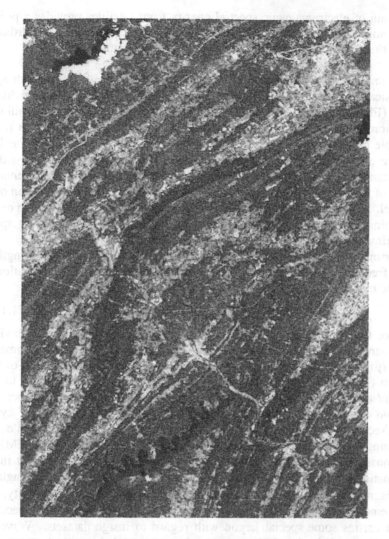

Fig. 1.1 Gray-tone image of band 2 (red light) from MSS sensor of Landsat satellite showing wooded ridges (darker) and agricultural valleys (lighter) characteristic of the Ridge and Valley Physiographic Province of central PA. Note clouds in upper left (northwest) corner, and Raystown Lake reservoir in the southern portion.

sensing of electromagnetic spectral energy from satellites provides a prolific source of environmentally relevant image data, we delve a bit further into this informational basis for imaging before enlarging our scope of signals.

Remote sensing gathers environmental information without establishing physical contact. It uses the spectral continuum of electromagnetic radiation (EMR) as a major source of signals. Visible light comprises a familiar portion of this continuum, with color being an important property in the visible range. EMR is a dynamic form of energy that propagates through space and various media, becoming evident only by its interaction with substances. It produces dual electric and magnetic fields with the orientation of these fields being perpendicular to each other and to the direction of travel. It propagates as a quantized wave, with the rate of propagation depending on the medium according to the *index of refraction* and being approximately 3×10^8 meters/second for a vacuum.

Primary properties of the EMR continuum (spectrum) are wavelength and frequency, which are related in terms of fixed velocity of propagation in the medium (see Eq. 1.1).

$$\text{Velocity} = \text{wavelength} \times \text{frequency.} \qquad (1.1)$$

For the spectral ranges of most common interest with regard to environmental remote sensing, wavelength is typically expressed in micrometers (μm) or fractions thereof, with a micrometer being one millionth of a meter (1×10^{-6} m). The basic unit of frequency is the hertz, which is one cycle/second. Corresponding to wavelengths in μm, spectral frequencies would commonly be expressed in gigahertz (1×10^9) or billions of cycles/second. This frequency unit has gained some public familiarity in describing clock cycle speeds of personal computers. Since velocity of EMR is constant for a medium, wavelength and frequency carry equivalent information. In the parlance of environmental remote sensing, wavelength (λ) is more commonly used than frequency to express spectral specificity.

Remote sensing of EMR by special detectors aboard aircraft and spacecraft carries some special jargon with regard to image datasets. Wavelength intervals having different properties are recorded separately as multiple spectral bands, which are sometimes also called *channels* in remote sensing.

Visible light occupies the approximate wavelength range from 0.4 μm to 0.7 μm. Within this spectral range, color changes with wavelength. White light is a balanced mixture of colors across this range. The general wavelength ranges corresponding to different colors are:

Violet =   .40 – .45 μm

Blue =	.45 – .50 μm
Green =	.50 – .57 μm
Yellow =	.57 – .59 μm
Orange =	.59 – .61 μm
Red =	.61 – .70 μm

In a rough manner of speaking, wavelengths from 0.4 to 0.5 μm can be said to be bluish, 0.5 to 0.6 greenish, and 0.6 to 0.7 reddish. Wavelengths shorter than 0.4 μm are not visible to the human eye, and are said to be *ultraviolet* (UV). Wavelengths longer than 0.7 μm are also not visible to the human eye, and are said to be *infrared* (IR).

Humans actually have a synthetic tricolor perception, with other colors being induced by mixed response of red-sense, green-sense, and blue-sense receptors. For example, yellow is perceived when red-sense and green-sense receptors are stimulated equally with no contribution from blue-sense receptors. Thus, wavelengths in the yellow range are effective in stimulating two of the three kinds of receptors. White or shades of gray are perceived when all three kinds of receptors are stimulated in a balanced manner. This is the basis for generating many different colors on computer screens using only red, green and blue (RGB) elements to be explained in some detail later.

EMR wave energy comes in specific quantities (quanta) according to wavelength, with shorter wavelengths carrying more energy. This relationship can be stated in terms of frequency, where Q is the quantity or flux of energy and (6.62×10^{-34} joule sec) is called Planck's constant. Cosmic rays and X-rays are very short wavelengths that carry sufficient energy to damage biological tissue by ionization (see Eq. 1.2).

$$Q = (6.62 \times 10^{-34} \text{ joule sec}) \times \text{frequency} \qquad (1.2)$$

In contrast, radio waves are very long wavelength and low energy forms of EMR. Short wavelengths are emitted by substances at very high temperatures such as the gases in the sun, whereas long wavelengths are emitted by substances at lower temperatures such as the thermal radiation of substances in the earth environment. As the temperature of a substance increases, it's wavelength of maximum emission decreases according to Wien's Displacement Law where T is (absolute) temperature in degrees Kelvin. (see Eq. 1.3).

$$\lambda_{max} = 2898/T \qquad (1.3)$$

The relationship between emitted energy flux and temperature is given by the Stefan-Boltzman Law where *k* is a constant and **T** is again absolute temperature in degrees Kelvin (see Eq. 1.4). Thus, the intensity of radiant energy emitted increases exponentially as the fourth power of the temperature of a substance.

$$F = kT^4 \qquad (1.4)$$

When radiant energy of a specific wavelength encounters a particular kind of substance, interaction or lack thereof is in three ways. Some fraction can be *absorbed* and contributes to raising the temperature of the substance, some fraction can be *reflected* to bounce off in a different direction, and the rest (if any) is *transmitted* through the substance according to the index of refraction. Because of reflection, refraction, and diffraction, the energy that is not absorbed is typically *scattered* in different directions. The partitioning of the incoming radiant energy among absorption, reflection, and transmission differs according to wavelength of EMR and the nature of the substance. It is the differential characteristics of these interactions that lend utility to remote sensing for environmental analysis.

The short wavelengths of ultraviolet energy interact very strongly with the gaseous, particulate and water vapor components of the atmosphere to the extent that rays coming from materials at or near the surface of the earth are trapped or dissipated in many directions before reaching remote sensing devices on high flying aircraft or orbiting spacecraft. Because of this *attenuation*, there is relatively little opportunity to use the UV range of wavelengths for environmental indicator purposes except from low altitudes.

The shorter wavelengths of visible light in the violet and blue color ranges are likewise strongly attenuated by atmospheric components to the degree that aerial and space images of the earth surface taken in this manner are typically quite hazy. Blue light has important environmental characteristics, however, by virtue of some ability to penetrate shallow clear water for showing bottom characteristics and being heavily absorbed by chlorophyll in the process of photosynthesis by plants. Rendering remotely sensed images in true color also requires recording of blue information; otherwise, pseudo-color (false-color) schemes must be used as explained later.

The importance of green light for environmental purposes should be evident from the fact that most plants are distinguished by their green color. This green color is due to absorption of blue and red by the plants for use in photosynthesis, with green being reflected (rejected). Thus, green is really the color that is useless to plants for purposes of photosynthesis. It should also be noted that other surfaces such as dry bare soil actually reflect more green light than plants. In the latter case, however, the green is a component of a spectral mixture that appears to the eye as a different color. The absorption of green light by water is substantial, but less than for longer wavelengths such as red and infrared.

Red light is important with respect to remotely sensed environmental indicators because it is strongly absorbed by plants for photosynthesis and has wavelength long enough that atmospheric interference is less than for blue and green. The red skies of morning and evening are a result of red light passing through the long atmospheric path of low sun angles while blue and green ranges are attenuated. Red light, however, is absorbed rather completely by water and even wet surfaces. The image in Fig. 1.1 was acquired in early September when bare fields appear lighter and forest foliage darker.

Invisible infrared radiation from about 0.7 to 1.1 μm wavelengths is extremely important for environmental remote sensing. Although it cannot be perceived with the human eye, it can be readily recorded with special sensors and films. This kind of radiation is very strongly reflected from the foliage of healthy broadleaf vegetation, and to a somewhat lesser degree from the foliage of healthy needle-leaved trees. This reflectance is even further informative by virtue of the fact that it drops drastically when the plants become unhealthy or are placed under severe stress from lack of moisture. There is essentially complete absorption of these wavelengths by water and wet surfaces except for clouds. This range of infrared wavelength is often called *near infrared* because the size of the wavelengths is near to the size of visible wavelengths. The range is nevertheless of somewhat longer span than that for the visible light, and some variation within it does occur. Therefore, the near infrared is often broken into several separate bands by multi-band sensors.

Several environmental remote sensing systems also record infrared spectral bands in the wavelength range 1.1 to 4.0 μm. This is a range of varied properties and spans the transition from mostly reflected solar energy to energy emitted by objects in the earth environment. Some parts of this range are also strongly attenuated by atmospheric constituents, particularly water vapor.

The only additional wavelength range that we directly consider here is very long wave infrared radiation called *thermal* energy comprised of heat waves. This range of wavelengths is usually considered to be approximately 10.4 to 12.6 μm. Cool objects are 'dark' in this range, whereas hot objects are 'bright'. Here the term 'brightness' is used to refer to the intensity of the radiant energy.

An important point of summary for this remote sensing introductory is the complementary nature of red and near infrared bands with respect to vegetation, as can be seen by comparing Fig. 1.2 (near infrared) with Fig.

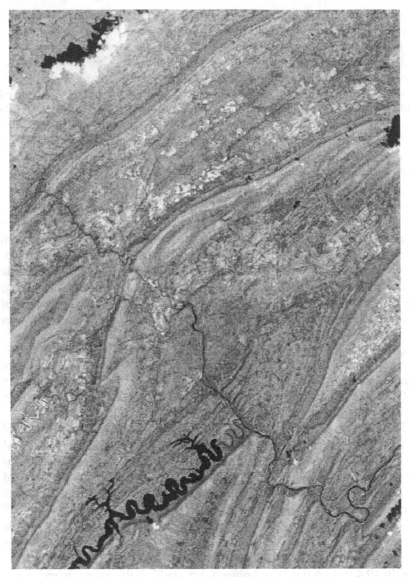

Fig. 1.2 Gray-tone image of band 3 (near-infrared 0.7-0.8 μm) from MSS sensor of Landsat satellite showing same area of central Pennsylvania as in Fig. 1.1.

1.1 (red). Red light is strongly absorbed by healthy vegetation, whereas near infrared light is strongly reflected by healthy vegetation. There is also more difference in reflectivity between broadleaf trees and needle-leaved trees in the infrared than there is in the red. Dry mineral soil tends to be bright in both bands, whereas water and wet areas are dark in both.

Insofar as possible, complementary signals such as red and near infrared should be chosen for multi-band images so that the combination carries more information than would be available by treating the bands individually. Conversely, using substantially redundant (highly correlated) signals for multiple bands will not confer much analytical advantage.

1.2 Satellite Sensing Scenario

In a general frame of reference for remote sensing, there is a primary distinction between *passive* remote sensing systems and *active* remote sensing systems. A passive system records radiant energy that is already present in the environment, such as reflected solar energy. An active system generates its own signal as a beam of energy, and then records the component of the beam that is reflected back to the sensor. The information from active systems such as *radar* and *lidar* requires special processing.

The current context for remote sensing typically involves a multispectral scanner containing several detectors, with the different detectors recording EMR for specific spectral bands. The scanner has an *instantaneous field of view* (IFOV) that sweeps across the landscape in lines, with records of EMR intensity being made at regular intervals along each line. Each such record for the set of bands constitutes a pixel for the image data. The bands are usually, but not always, ordered according to wavelength from shorter wavelengths to longer wavelengths. Although the actual mechanics of the process varies considerably among sensing systems, the result is basically as outlined.

There are numerous remote sensing systems that differ in spatial resolution, spectral bands, and frequency of image acquisition. The types of systems available also change over time. Cataloging the characteristics of these various systems is not the present purpose, but the nature of sensors in the NASA Landsat series is central to explaining and illustrating applications of the analytical approach. Landsat data represent the earliest publicly available space-based multiband environmental information, and therefore constitute the basis for analyzing longer time spans of landscape change dating back to 1972.

Fig. 1.6 Synthetic image showing distribution of fish species in Pennsylvania according to regional habitat importance index (RHII), with darker tones indicating more important habitats.

and modeling in ecology and environmental science, and also serve here as examples of subsequent analysis. When monitoring entails sampling that is distributed across a region, geostatistical interpolation can be used to obtain a synoptic expression of the information as a specified spatial lattice (Schabenberger & Gotway, 2005).

1.4 Georeferencing and Formatting Image Data

In order for image data to be coupled analytically with other data, it is necessary to have capability for locating the geographic position of any particular pixel. There are also several possible protocols for encoding the signal information in computer files. These are typically large data sets, often containing millions of values.

With respect to spatial position, column number counted from the left serves as a relative X-coordinate and row number counted from the top serves as an inverted relative Y-coordinate. If the side dimension of a pixel is known in absolute coordinate units along with the X and Y location of the upper left corner, then location of a pixel center can be computed with the absolute units increasing upward on the Y-axis. Thus, the direction of increase for the Y-axis of absolute units is opposite that of row numbers (see Eqs. 1.6 and 1.7).

$$X = \text{starting } X + (\text{column number} - 0.5) \times \text{pixel dimension} \qquad (1.6)$$

$$Y = \text{starting } Y - (\text{row number} - 0.5) \times \text{pixel dimension} \qquad (1.7)$$

Storing row numbers and column numbers explicitly, however, would considerably increase the amount of computer media required. Therefore, a sequencing convention is adopted for arranging information in the file that implicitly allows translation of position in the file to spatial position and vice versa. Information is stored generically in row major order, which works from top (first row) to bottom and from left (first column) to right. For a lattice containing only one value (signal band or property) per pixel, the order of storage is fully determined by this convention.

For a lattice containing several values (signal bands or properties) per pixel, however, the convention must also specify how values are sequenced among bands. The convention that we adopt for present purposes is to have the values of the respective bands interleaved by pixel (BIP), which is to say that all of the band values for a pixel are given in order before proceeding to the next pixel in the row. With this convention, the number of values that must be skipped in a file to reach the start of information for the pixel in row r and column c is given by Eq. 1.8.

$$\text{bands} \times (r-1) \times (\text{columns per row}) + (c-1) \times \text{bands} \qquad (1.8)$$

Another common arrangement of band values is to have the values for the first band in sequence across the entire first row, then the values for the second band in sequence for this row, and so on to the last band for the same row. The sequencing then proceeds in like manner for the second row, the third row, etc. This method of sequencing is said to be band interleaved by line (BIL).

Still another way is to have row major entry of the first band for all pixels in all rows, followed by row major entry of the entire set of values for the second band, and likewise for the third and subsequent bands. This third method of entry is said to be band sequential (BSQ).

A further important aspect of image-structured data is the manner in which band values are encoded on computer media. Because of the multitude of band values involved, it is common to devote only a minimal amount of media to each value. The most efficient allocation of media in this respect is a *byte* comprised of eight *bits*, with each bit being capable of representing two states. This enables representation of 256 states in a byte, with the states conventionally signifying integer numbers from 0 to 255. If two bytes are allocated to each value, then the range of integer representation becomes 0 to 65,535. With two bytes there are further alternatives for starting with the low portion of the number in the first byte or the high portion, and for cutting the range in half to represent both positive and negative integers.

It should be apparent that the several bands of image data must be spatially matched or 'registered'. This may require reconstructing a lattice for each individual band in the process of assembly as a multiband dataset. Such reconfiguration is an advanced interpolative operation called *resampling* that is best accomplished with commercial software packages for image analysis. The general strategy is to determine for each pixel of the new lattice which value should be used from the old lattice. One expedient way of doing this is to take the value from the (geographically) nearest old pixel, which is called the 'nearest neighbor' approach. Another way is to interpolate among nearby pixels, with several interpolation protocols being available. Interpolation requires that the values in the lattice have a well-ordered gradation of measurement (not categorical). There are several circumstances in which data may be missing for a portion of the lattice, which requires that there be a code assigned as a missing data signal. Zero usually serves the purpose of a missing data flag. For our purposes, a pixel that has no data for any layer must be considered as having no data for all layers.

There is a considerable variety of software systems for spatial analysis, both commercial and otherwise, that can accept some variant of these generic data structures as input. However, any given software system may restructure the data internally for compatibility with its own particular method of data management. The auxiliary information on size of the grid, number of layers, sequencing, encoding, and location is typically kept in a small 'header' file separate from the data itself. This header file is often readable with a text editor, but not necessarily so.

1.5 The 4CS Pattern Perspective On Image Modeling

Our formalizing of patterns for image modeling is intended to serve several purposes in relation to images of landscapes. These include data compression as well as analytical advantage and extended enhancement of pictorial presentation. Fig. 1.7 offers a framework for capturing, conceptualizing and visualizing the multipurpose nature of the approach that we are pursuing through pattern processes for multiband images of landscapes. An aspect to be emphasized in the conceptualization is that images embody (spectral) *signal*, *spatial* and *sequential* (temporal) infor-

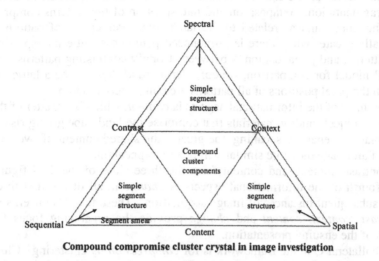

Compound compromise cluster crystal in image investigation

Fig. 1.7 The 4CS framework of image modeling by segmentation with compressive pattern processes.

mation concerning landscapes. The fourth S in the 4CS framework concerns the segmentation of an image that arises naturally from any multi-pixel patterns that it contains. Each pattern component constitutes a sub-image or *segment* of the image (Gonzalez & Woods, 2002).

Modification of the primitive patterns to make them more prominent by an aggregating (clustering) pattern process will induce some blurring or smear of detail in the original image. Therefore, strengthening of patterns entails some compromise relative to image detail. The compromise is entirely inconsequential only when the suppression of primitive patterns is limited to (usually electronic) noise that is spurious relative to the intended signal. This smearing is symbolized by the outer frame of the 4CS figure.

The informational aspects of the image can also be seen as being four-fold. The first of these four aspects is *contrast*, which involves comparative differences between patterns of the image that are perceived visually as light versus dark in gray-tones or as differences in color. Increasing the contrast will strengthen the perception of at least some of the patterns in the image. Clouds and their shadows are examples of high contrast features of environmental images, albeit often being nuisance patterns for analysis of the landscapes they obscure.

The second of the four information aspects is *context*, which involves the fragmentation, juxtaposition and interspersion of the patterns comprising the image and as related to the landscape. Perception of pattern is typically greater when there is pronounced patchiness, little interspersion of patterns, and juxtaposition is between strongly contrasting patterns. As a null model for comparison, we consider a *'randscape'* to be a lattice in which the pixel positions of all patterns are completely random.

The third of the informational aspects is *content*, which is a matter of the kinds of objects and/or materials that comprise the landscape giving rise to the image. Generally speaking, the greater the shared content of two sectors of an image the more similar will be their appearance.

Contrast, context, and content form the three sides of the 4CS figure. The fourth of the informational aspects is *change* in all of the first three with subsequent instances of image acquisition. These four C concerns of *contrast, context, content* and *change* provide the organizing focus for much of the ensuing presentation.

A collateral C in the framework is for *compression* by clustering. Clustering is the statistical term for methods that segregate dissimilar entities from similar entities or conversely for aggregation as opposed to segregation (Mirkin, 2005). To the extent that differences among the members of a cluster are ignored, the clustering leads to some loss of detail in the information. On the other hand, clustering provides opportunities for more parsimonious recording of the information to the degree that cluster labels can be used to record occurrence and the properties of the cluster can be recorded in a table for which the cluster label serves as an index entry. In short, a vector of cluster properties need be recorded only once in tabular form with cluster labels used elsewhere for cross-reference to the table. This leads to reduced volume of data which constitutes a form of *compression* according to the jargon of image analysis (Gonzalez & Woods, 2002). To avoid possible confusion with statistical clusters, we refer to spatially connective aggregations of pattern elements as *patches*, *instances* or *connected components*.

As an innovative part of our extended treatment of patterns, we introduce here a way to have strong pattern 'overtones' in an image model while reducing the attendant loss of 'undertone' detail. This is by multi-level indexing of patterns to form nested patterns. Thus, we can have a first (strong) level of numbered patterns along with a table of *proxy* signal vectors expressing central tendencies for those strong patterns. Then each of the first-level patterns can be disaggregated into a numbered series of second-level patterns, each having another more specific vector of signal properties that differentiates it from others having like index and vector at the first level. We use the term *polypattern* to describe such compound patterns. The interior triangle of the 4CS framework symbolizes such compound patterns for multi-level image models.

References

Argent, D., J. Bishop, J. Stauffer, R. Carline and W. Myers. 2003. Predicting Freshwater Fish Distributions Using Landscape-Level Variables. Fisheries Research 60: 17-32.

Bolstad, P. 2005. GIS Fundamentals – A First Text on Geographic Information Systems. White Bear Lake, MN: Eider Press. 411 p.

Chrisman, N. 2002. Exploring Geographic Information Systems. New York: John Wiley & Sons. 305 p.

Cracknell, A. and L. Hayes. 1991. Introduction to Remote Sensing. New York: Taylor and Francis. 303 p.

Davis, F. W., D. M. Stoms, J. E. Estes, J. Scepan and J. M. Scott. 1990. An Information Systems Approach to Biological Diversity. *International Journal of Geographic Information Systems* 4: 55-78.

DeMers, M. 2000. Fundamentals of GIS, 2nd ed. New York: John Wiley & Sons. 498 p.

Frohn, R. C. 1998. Remote Sensing for Landscape Ecology: New Metric Indicators for Monitoring, Modeling, and Assessment of Ecosystems. Boca Raton, FL: Lewis Publishers. 99 p.

Forman, R. T. T. 1995. Land Mosaics: the Ecology of Landscapes and Regions. Cambridge, U.K.: Cambridge Univ. Press. 632 p.

Forman, R. T. T. and M. Godron. 1986. Landscape Ecology. New York: John Wiley & Sons. 619 p.

Gibson, P. and C. Power. 2000. Introductory Remote Sensing: Principles and Practices. New York: Taylor and Francis. 208 p.

Gonzalez, R. and R. Woods. 2002. Digital image processing, 2nd ed. Upper Saddle River, NJ: Prentice-Hall. 793 p.

Jensen, J. R. 2000. Remote Sensing of the Environment: An Earth Resource Perspective. Upper Saddle River, NJ: Prentice-Hall. 544 p.

Lillesand, T. and R. Kiefer. 1999. Remote Sensing and Image Interpretation 3rd ed. New York: Wiley. 750 p.

McGarigal, K. and B. Marks. 1995. FRAGSTATS: Spatial Pattern Analysis Program for Quantifying Landscape Structure. General Technical Report PNW 351, U.S. Forest Service, Pacific Northwest Research Station. 122 p.

Mirkin, B. 2005. Clustering for Data Mining. Boca Raton, FL: Chapman & Hall/CRC, Taylor & Francis. 266 p.

Myers, W., J. Bishop, R. Brooks, T. O'Connell, D. Argent, G. Storm, J. Stauffer, Jr. and R. Carline. 2000. The Pennsylvania GAP Analysis Final Report. The Pennsylvania State University, Univ. Park, PA 16802.

Myers, W., J. Bishop, R. Brooks and G. P. Patil. 2001. Composite Spatial Indexing of Regional Habitat Importance. *Community Ecology* 2(2): 213-220.

Myers, W., M. McKenney-Easterling, K. Hychka, B. Griscom, J. Bishop, A. Bayard, G. Rocco, R. Brooks, G. Constantz, G.P. Patil and C. Taillie. 2006. Contextual Clustering for Configuring Collaborative Conservation of Watersheds in the Mid-Atlantic Highlands. *Environmental and Ecological Statistics* 13(4). In press.

Myers, W, G.P. Patil, C. Taillie and D. Walrath. 2003. Synoptic Environmental Indicators as Image Analogs for Landscape Analysis. *Community Ecology*, 4(2): 205-217.

Olivieri, S. T. and E. H. Backus. 1992. Geographic Information Systems (GIS) Applications in Biological Conservation. *Biology International* 25: 10-16.

Schbenberger, O. and C. Gotway. 2005. Statistical Methods for Spatial Data Analysis. New York: Chapman & Hall/CRC. 488 p.

Scott, J. M., F. Davis, B. Custi, R. Noss, B. Butterfield, C. Groves, H. Anderson, S. Caicco, F. D'Erchia, T. C. Edwards, Jr., J. Ulliman and R. G. Wright. 1993. Gap Analysis: A Geographic Approach to Protection of Biological Diversity. *Wildlife Monographs* No. 123.

Turner, M., R. Gardner and R. O'Neill. 2001. Landscape Ecology in Theory and Practice: Pattern and Process. New York: Springer-Verlag, Inc. 401 p.

Walsh, S. and K. Crews-Meyer. 2002. Remote Sensing and GIS Applications for Linking People, Place, and Policy. Boston, MA: Kluwer Academic Publishers.

Wilkie, D. S. and J. T. Finn. 1996. Remote Sensing Imagery for Natural Resources Monitoring: A Guide for First-Time Users. New York: Columbia University Press. 295 p.

2 Pattern Progressions and Segmentation Sequences for IMAGE Intensity Modeling and Grouped Enhancement

The foregoing background regarding multiband digital image data provides a basis for pursuing our primary focus on patterns in landscape images. Landscapes are characterized by spatial autocorrelation (Schabenberger & Gotway, 2005) whereby things closer together tend to appear more alike than things that are further apart with some changes being gradational and others abrupt, which induces implicit perception of pattern. However, the indefiniteness of implicit pattern perception limits its utility. In landscape ecology, pattern has been most often addressed in terms of variously defined mosaics and parameters of patchiness (Forman & Godron, 1986; Forman, 1995; McGarigal & Marks, 1995; Turner, Gardner & O'Neill, 2001). Furthermore, pattern is a much used but rather varied conceptual construct for image analysis, as witness the lineage of literature relating to 'pattern recognition' including a journal so named along with disparate sources (Tou & Gonzales, 1974; Pavlidis, 1977; Gonzales & Thomason, 1978; Fu, 1982; Simon, 1986; Pao, 1989; Jain, Duin & Mao, 2000; Duda, Hart & Stork, 2001; Webb, 2002) and extensions into the contemporary topics of clustering, classification, machine learning, data mining and knowledge discovery. Therefore, an obvious next task is to resolve some of the indefiniteness regarding pattern in the current context. As a point of departure, we take the succinct statement of Luger (2002) that pattern recognition is identifying structure in data.

2.1 Pattern Process, Progression, Prominence and Potentials

We designate L as the set of pixel positions comprising the image lattice. Let us further reference the pixel position at row i and column j as being $L(i,j)$. We likewise designate V as the entire set of different signal vectors. Each of these vectors can be considered as a point in the space of signal properties, which we will call signal domain. Let us further reference a particular one of these *property points* as being pp_k for the kth such point in some fixed order. It should be intuitively apparent that a notion of landscape pattern must be a joint construct in the spatial domain L and the signal domain V. Accordingly, we define a *pixel pattern* or *pattern of pixels* PP_k as being the subset of pixels sharing the same property point pp_k (signal vector) and indexed by the index of the property point. We proceed to consider a process that assigns property points to pixels forming an initial image as being a *pixel process* symbolized by a superscript # resulting in a set $PP^\#$ of *primitive patterns*. Image data for landscapes tend to have large numbers of primitive patterns each of which encompasses relatively few pixels with considerable fragmentation and interspersion. Part of this multiplicity arises from 'edge effects' where pixels span boundaries in the landscape. Thus the primitive patterns tend to be visually subtle or weak. Our intent is to strengthen these patterns so that they become less numerous but more apparent and better segregated.

Toward this end, we now define a *pattern process* as one that produces sets or subsets from *prior patterns* to yield either more general or more specific *posterior patterns*. All of the pixels in a particular posterior pattern will share the same property point, which serves as *proxy* for whatever property point was applicable in a prior pattern. If a pattern process operates recursively, we call it a *progressive pattern process* and refer to its recursive sequence of pattern sets as a *pattern progression*. Since it operates across the image lattice L, we use the symbol £ to denote a pattern process.

An important aspect of pattern processes is the strength of the patterns that they produce. In order to make these concepts operational, however, measures of strength for patterns are needed. As a simple expression of pattern strength, we define *prominence of a pattern* to be the proportion p_k of non-null pixels for the kth pattern in the lattice. This is one formulation of what we will call a *mass function* $M(k)$ for the kth pattern.

Continuing with measures of strength for patterns, we emulate physics with analogies to ideas of potentials in fields of charged particles. Let the *potential of a pattern* or simply *pattern potential*, symbolized as Pp, have the form of Eq. 2.1

$$Pp(k) = [1 + a_1 M(k) + a_2 M(n_k)] \, D^z_w(k, n_k) \qquad (2.1)$$

where

$M(k)$ is a mass function for the kth pattern;

D_w is a (weighted) distance function between property points in signal space;

a_1 and a_2 are 'aggregation' parameters;

z is a 'zonal' parameter;

n_k is the nearest neighbor of property point pp_k by D_w in the signal domain.

Note that $Pp(k)$ pertains to a definite pattern PP_k relating a particular property point pp_k to a particular set of pixel positions. In parallel manner and with some appeal to the dual ideas of definite and indefinite integrals, we define a *potential pattern* $pP(k)$ pertaining to a property point that is decoupled from any particular positions using the same formula as for $Pp(k)$ but with a mass function that does not explicitly reference positions in the lattice. *Intrinsic potential* $pP'(k)$ is obtained by setting the aggregation parameters to zero, which inherently decouples property points from pixel positions since the distance is measured in the signal domain.

The aggregation parameters and mass functions have an effect of local dilation or expansion of signal space. Two patterns effectively have greater separation with increase of any parameter. Even when the aggregation parameters are equal, a pattern does not necessarily have the same potential as its nearest neighbor because that neighbor may have a different nearest neighbor.

2.2 Polypatterns

In order to strengthen subtle patterns without suppressing them completely, we can work with patterns of patterns as compound patterns or *polypatterns*. This entails multi-level indexing of patterns to form nested patterns. Thus, we can have a first (strong) level of numbered patterns along with a table of property vectors (points) for those strong patterns. Then each of the first-level patterns can be disaggregated into a numbered series of second level patterns, each having another more specific property vector that differentiates it from others having like index and vector at the first level. We designate the aggregated level of a bi-level pattern as the A-level, and the disaggregated level as the B-level (base level). Property vectors in the A-level serve as *proxies* for property vectors in the B-level. To minimize further complexity of notation, polypatterns can be symbolized by prefacing an entire pattern reference with the ‡ (double dagger) symbol in the notation previously defined where clarification is required. Thus, $‡p_k$ symbolizes the prominence of the kth polypattern.

In constructing index numbers as identifiers for polypatterns, the second level of numbering can either run across the first levels globally or be conditional within the first level. The global approach is more convenient for accessing look-up tables of the more detailed (B-level) pattern properties,

but the larger sizes of the index values involved also occupies more computer media and thus reduces the degree of compression.

As a compromise, we use conditional numbering but also record the
cumulative number of finer segments by pattern number in the coarser
level. Let m_k be the number of disaggregated patterns in the kth polypattern, and let these patterns be conditionally numbered $1...m_k$. As a globally sequential index $g_{k,i}$ to the ith disaggregated (B-level) pattern in the kth
polypattern we use Eq. 2.2.

$$g_{k,i} = i + \sum m_{j<k} \qquad (2.2)$$

This index can be computed dynamically from the A-level polypattern
numbers and the conditionally (nested) numbers of the disaggregated patterns coupled with a distributional list of pattern frequencies for the A-
level polypatterns. The only overhead of computer media relative to compression is for storing the distributional list of pattern frequencies for A-
level polypatterns.

The creation of polypatterns allows for exploitation of the ideas underlying 'constrictive analysis' as described by Myers, Patil and Taillie (2001).
Polypatterns should also help to clarify our 'proxy' terminology for property vectors associated with pattern processes. A property vector at the
aggregated A-level serves as a proxy for all of the patterns encompassed at
the disaggregated B-level.

2.3 Pattern Pictures, Ordered Overtones and Mosaic Models of Images

A set of PP patterns forms a spatial mosaic on the image lattice, regardless
of whether these are primitive patterns, clustered (proxy) patterns, or polypatterns. To the degree that the signal values in the pp property points of
the patterns may be reasonable proxies for the original pixel vectors, it
should be possible to treat a mosaic of patterns as an approximate model of
the image data from which they were obtained. Likewise, a *pattern picture* resembling the original image should be obtainable by using selected
elements of the pattern properties to portray the patterns as tones in a mapping mode.

Such graphic image emulation is facilitated by making it possible to
treat an A-level pattern mosaic like a simple single band of image data for
direct display.

An image-like mosaic can be constructed by making the pattern index
numbers correspond with the ordering of overall intensity using the signal

bands comprising the pattern vectors (property points). The patterns having lesser overall intensity for their signal bands are given lower index numbers, and those having greater overall intensity are given higher index numbers. In other words, the patterns are ranked according to overall (or average) intensity and the rank number becomes the identifying index number for the pattern. The pattern index numbers are entered directly in an image lattice. The pattern numbers thus become an index image of overall signal intensity which can be treated as brightness in a gray-tone image. The norm (or length) of the signal vector as distance of the property point from the origin of signal space is one convenient measure of overall intensity, computed as the square root of the sum of squared signal values and being a multiple of the quadratic mean of the values across signal bands. *Ordered overtones* appear in a pattern picture that translates the pattern identification numbers directly to gray tones.

Ordered overtones derived from ASTER sensor data for central Pennsylvania are shown in Fig. 2.1 as produced by pattern processes described subsequently. The ASTER acronym is for Advanced Spaceborne Thermal Emission and Reflection Radiometer on a Terra satellite operated by NASA with Japanese cooperation. The image data used as a basis for Fig. 2.1 were acquired by the sensors on September 6, 2002. Ten ASTER bands were harmonized in 15-meter pixels, with the central wavelengths of the spectral bands being as shown in Table 2.1.

Table 2.1 Center wavelength (μm) of ASTER bands used for Fig. 2.1.

Band Number	Center wavelength (μm)
1	0.556
2	0.661
3	0.807
4	1.656
5	2.167
6	8.291
7	8.634
8	9.075
9	10.657
10	11.318

Fig. 2.1 Ordered overtones of ASTER satellite data for central Pennsylvania, September 2002.

Similarly, Fig. 2.2 shows ordered overtones of the six-band vertebrate species richness data for Pennsylvania developed in like manner. Since signal intensity corresponds to increasing species richness, the lighter (brighter) areas in Fig. 2.2 exhibit greater species (habitat) richness as averaged over the taxonomic groups being treated in the manner of signal bands as described in Chap. 1.

Despite this quasi-quantitative treatment of identifying indexes for polypatterns, we should not lose sight of the fact that our patterns are inherently categorical constructs. The properties of a pattern are quantitative, but a pattern is a categorical collective.

Fig. 2.2 Ordered overtones of habitat richness patterns in Pennsylvania for six groups of vertebrate species being treated as signal bands with lighter as greater.

2.4 Pattern Processes for Image Compression by Mosaic Modeling

The foregoing background and definitions lay the groundwork for using segmentation sequences as progressive pattern processes for image compression by mosaic modeling for purposes of landscape analysis as explored in subsequent chapters. The overall goal is a parsing of patterns in which the proxy properties for the patterns closely approximate the pixel primitives of the image while being few enough to record using substantially less electronic media than occupied by the original image data. In a general sense, this is a problem of strategically segmenting the image. In a statistical sense, this falls most directly under the subject of cluster analysis. However, our approach involves configurations and combinations of processes that are not conventional with regard to either clustering or image analysis.

There are four important aspects of the undertaking. One is to strengthen the primary patterns in the image so that landscape structure becomes more evident. A second is to retain substantial information regarding subtle patterns in the image, particularly so as not to induce extensive areas that are lacking in any detail. A third is to achieve a substantial degree of compression within these constraints. Fourth is to accomplish all of these in a manner that is computationally practical for large images, with this latter being somewhat dependent on the computer configuration. Therefore, compromise is inherent – which favors heuristics over optimization.

The first and second aspects can be accommodated through bi-level polypatterns, in which the stronger primary patterns comprise an A-level and more subtle pattern variants comprise a B-level. Even with polypatterns, the fourfold problem is excessively open-ended; which can be remedied somewhat by adding a fifth aspect of having the primary patterns be compatible with geographic information systems (Burrough & McDonnell, 1997; Chrisman, 2002; DeMers, 2000). The desired compatibility can be achieved by allotting one byte per pixel as a GIS raster (cellular grid) layer. The layer structure for primary patterns is then complemented by allotting a second byte per pixel for the more subtle pattern components in the B-level of polypatterns. Since it is common practice to record image data with one byte per band in each pixel, this structure essentially occupies the equivalent of two bands of image data. The degree of compression afforded by the fixed layer structure of polypatterns will thus depend on the number of bands, being inapplicable for fewer than three bands. The fixed layer structure is also amenable to further content-based compression, provided that the requisite layer arrangement is restored by decompression prior to usage.

One byte affords a possible 256 pattern distinctions in the A-layer. However, zero is needed as a designator for pixels in the lattice that do not pertain to the image area of interest – thus leaving a possible 255 pattern distinctions. A strategic decision has been made to reserve five designators for special usage in GIS mapping, thus providing for 250 primary patterns in the A-layer. Each of the 250 A-level patterns can have 255 B-level sub-patterns, therefore allowing for a total of 250×255=63,750 patterns. The usual number of patterns will be at least an order of magnitude less than this due to computational constraints and unequal distribution of subtle variants among the 250 primary patterns.

There are two avenues of analysis toward this sort of polypattern parsimony, and both involve several stages of image segmentation and/or clustering. One is to segregate 250 segments in the early stages, and then segment the segments in later stages. The other is fine segmentation in the early stages, and then later stage aggregation of fine segments as primary patterns. Other avenues of alternation in segregations and aggregations are also possible. Our approaches entail four phases.

Since these are compound pattern process scenarios with patterns representing image segments, the idea of a *segmentation sequence* becomes useful. Aside from primitive pixel patterns, the number and type of patterns arising from a particular segmentation stage becomes important. Therefore, the notation of Eq. 2.3 for a segmentation sequence is used to symbolize a scenario $£_\alpha$ that entails four stages producing 250, 250, 2500 and 250 patterns, respectively, with the kinds of patterns as explained below.

$$£_\alpha\{\#250?|\#250*|\#2500*|*250\ddagger\} \tag{2.3}$$

As defined previously, the symbol $£$ is generic for a pattern process. The subscript α indicates that this is the particular 'alpha' segmentation sequence of pattern processes. The numbers in curly brackets are the numbers of patterns produced by the respective stages. The vertical bar | symbol separates the stages comprising the sequence. The leading and trailing symbols for the numbers indicate the kind of pattern that the stage operates upon and the kind of pattern that it produces, respectively. The # symbol indicates a primitive pattern, the ? symbol indicates a potential pattern, the * symbol indicates a proxy pattern, and the ‡ symbol indicates a polypattern. Thus the first stage of this sequence operates on primitive patterns and produces 250 potential patterns; the second stage operates on primitive patterns and produces 250 proxy patterns; the third operates on primitive patterns and produces 2500 proxy patterns; the fourth operates on proxy patterns and produces 250 polypatterns.

2.5 α-Scenario Starting Stages

Our initial analytical approach to pattern parsing is an α-scenario that has been extensively tested on images and provides the point of departure for subsequent scenarios. It consists of a four-stage segmentation sequence as addressed above.

This produces bi-level polypatterns that are mapped into two bytes for each pixel as two 1-byte lattices with auxiliary tables. The aggregated A-level of patterns is arranged as ordered overtones that can be rendered directly as a gray-scale image. The (finer) B-level of the polypatterns must be decompressed with custom software to obtain more generic data for input to other software packages having facilities for image analysis.

The first stage of the α-scenario takes primitive patterns from multi-band image data in byte-binary BIP format and produces 250 potential pat-

terns as output. This process is based on intrinsic potentials for property
points using the square of (weighted) Euclidean distance (zonal parameter
z=2) as Eq. 2.4.

$$pP'(k) = D^2_w(k,n_k) \tag{2.4}$$

The first 250 non-duplicate pixels in the data file are taken as the initial
potential patterns, among which the weakest potential will be shared in a
pair. Each subsequent pixel is considered with regard to intrinsic potential
among the current set, displacing the first of the weakest pair if it is
stronger. Since the signal vectors of the pixels are considered conditionally on their order of occurrence in the file, this does not necessarily yield
the strongest set of potential patterns; however, it does give a strong set
encompassing property points that are well distributed in signal space.
Unequal weighting of the signal bands for computation of squared distance
is optional, with the defaults being unit weights.

The second stage of the α-scenario scans the primitive patterns from the
same multiband image data and produces pixel patterns (PP) by associat-
ing each pixel with the potential pattern to which it is closest by (weighted)
Euclidian distance in signal space (i.e., closest property point). Thus the
potential patterns from the first stage become the proxies for the pixel patterns from the second stage. These pixel patterns are then ranked according to norm of signal vector for overtone in concluding the second stage.

2.6 α-Scenario Splitting Stage

The third stage of the α-scenario partially disaggregates the (proxy) pixel
patterns from the second stage, but works with primitive patterns from the
original image in doing so. Disaggregating is accomplished by recursive
(unequal) bifurcations, thus making a tree of (binary) branching nodes for
those patterns that are subject to splitting. Not all of the pixel patterns
from the second stage, however, are subject to splitting. Computational
constraints also impose practical limits on the number of (nodal) bifurcations that can be conducted concurrently. Therefore, (tunable) parameters
of practicality affect the course of the disaggregating process. The set of
(primitive) patterns for a node of the tree constitutes a *constellation* of
property points in signal space.

The fundamental mechanism of splitting in the α-scenario is one of polar proximity according to Euclidean distance. For each node, the signal
vector having the largest norm (most intense or 'brightest') and smallest
norm (least intense or 'darkest') are determined as poles of an axis span-

ning the constellation of property points. When bifurcation takes place, those primitive patterns closest to each of the two poles become pattern subsets; and the poles of the subsets are determined in doing the segregation. A synthetic signal vector is calculated as the middle of the polar axis by averaging the respective components of bright and dark poles to serve as proxy for the new nodal subset. In short, the proxy point is at the midrange of intensities. Figure 2.3 is a diagrammatic representation of this splitting.

We extend the idea of potentials to obtain a prioritizing criterion Φ for splitting. This criterion, which we call *polar potential*, is computed for each node by Eq. 2.5 as the product of the inter-polar or axial (Euclidean) distance and the number of pixels comprising the node,

$$\Phi_{k,j} = (\Delta_{k,j})(\varphi_{k,j}) \tag{2.5}$$

where $\Delta_{k,j}$ is the axial or inter-polar distance for the jth node in the kth second-stage pattern, and $\varphi_{k,j}$ is the number of pixels in that nodal group (subset). The polar potential criterion thus reflects both prominence and diversity of the parent pattern.

Concurrent splitting is conducted for the H highest nodes as ordered on the polar potential criterion, where H is the computational capacity for concurrent splitting. A minimum threshold is also imposed on the criterion for a node to be eligible for splitting. The polar potentials are recomputed after each episode (cycle) of concurrent splitting, with new nodes being considered for the queue. Since the number of nodes can only increase by H in each cycle, there is effectively a competition for place in the splitting queue, with advantage going to prominence and diversity. The capacity H for concurrent splitting is one of the factors of the competition, with a higher capacity allowing smaller and less diverse nodes to be split. As a consequence, lower capacity for splitting gives greater depth in structure of the trees and requires more cycles to reach a given number of nodal patterns. An overall limit on number of nodes is also imposed so that the ultimate polypatterns can be coded in two bytes per pixel.

2.7 α-Scenario Shifting Stage

The fourth and final stage in the α-scenario is one of rearranging aggregations that operates on the nodal patterns of the third stage instead of upon primitive patterns of the original image data. This process (re)groups the nodal patterns of the previous stage into A-level polypatterns. Individual differences between primitive patterns in a nodal constellation are dis-

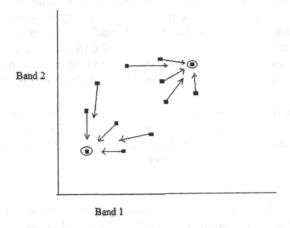

Band 2

Band 1

Fig. 2.3 Diagrammatic representation of polar partitioning. Small rectangles are property points for 2-band signal vectors. Enclosed property points are poles.

carded in formulating the final bi-level polypatterns. The primitive patterns of the original image can only be approximated with at least some error in at least some of the pixels. The polypatterns thus constitute a 'lossy' compression of the original image data (Gonzalez & Woods, 2002). The combination of pattern enhancement, compression, mapping and inherent inability for exact restoration makes the polypattern image intensity model a fundamentally different derivative product from the original image data in the same manner as a thematic map would be. Thus, copyright restrictions on redistribution of the original data should be obviated in most respects.

The fourth-stage (re)aggregation again entails compromise between competing practicalities. Overall information content for the A-level is favored by avoiding large discrepancies in prominence between the A-level patterns. This follows from information theory, but also from simply viewing each pattern as a carrier of information with a view to not having available carriers that are nearly empty. On the other hand, relatively restricted features of landscape pattern such as roads, streams and smaller water bodies tend to be defining features of terrain that should not be lost by blending with more ubiquitous pattern elements. By way of compromise, a target minimum prominence for aggregated patterns is a process parameter along with the obligatory maximum of 255 B-level members per A-level pattern for two-byte encoding. The target minimum prominence incorporated in the algorithm is p=0.00025, which is one-fortieth of one percent of the pixels.

The nodal trees for the second-stage A-level patterns provide an organizing template for the ultimate polypatterns. Preference is given to retaining such a tree as an A-level unit provided that it encompasses minimum prominence and does not exceed 255 nodes. If a tree does not attain minimum prominence, it is preferentially augmented by shifting nearest (by Euclidean distance) nodes from neighboring trees on a single-linkage (Podani, 2000; Mirkin, 2005) basis according to proxy signal vectors. Since the nodal splitting mechanism can segregate some very sparse patterns, a minimum prominence is also operative in this regard and similar single-linkage nodal shifts made accordingly. The final restructuring is to prune any remaining trees with more than 255 nodes backward from the terminals (by collapsing previous splits) to reach the 255 limit.

A segmentation sequence for the overall α-scenario takes the form of Eq. 2.6 with the actual number of third-stage patterns depending on the

$$\pounds_{\alpha}\{\#250?|\#250*|\#\max 63500*|*250\ddagger\} \qquad (2.6)$$

capacity for concurrent splitting and number of splitting cycles conducted. The algorithmic implementation allows splitting to be continued across several computing sessions.

The ordered overtone images in Figs. 2.1 and 2.2 were generated by this α-scenario. The segmentation sequence for Fig. 2.1 is:

$$\pounds_{\alpha}\{\#250?|\#250*|\#1898*|*250\ddagger\} \qquad (2.7)$$

and the segmentation sequence for Fig. 2.3 is:

$$\pounds_{\alpha}\{\#250?|\#250*|\#1248*|*250\ddagger\} \qquad (2.8)$$

Mirkin (2005) advocates capability for data recovery as a means of evaluating a method of clustering. Recovery of information along with mapping the spatial distribution of relative residuals for the α-scenario is considered later.

The α-scenario of polypattern processing is highly heuristic and was developed adaptively during the course of several years for different project purposes using a variety of image data sources, as well as being used extensively for instructional purposes in image analysis. The inception of that scenario was influenced by the early work of Kelly & White (1993). It has been consistently very robust in these various contexts. Recent

pairs, which is computationally impractical for present purposes. A more
expedient approach is to consider only pairs of poles that span the constellation in some sense.

The current sense of spanning is pairs of *peripheral points* according to
the following concept. Let signal vectors (property points in signal space)
P_1a and P_1b be taken as an initial polar pair, where P_1a has minimum norm
and P_1b has maximum norm. Then replace either member of that polar
pair by the most distant property point from its opposite pole (other than itself) as a new polar pair. Let each of a succession of such replacements
also give a polar pair. Then define the set of all possible members of such
polar pairs as being the set of peripheral property points (signal vectors).

The third stage of the α-scenario is based entirely on one such pair consisting of the points closest to and farthest from the origin of signal space
('darkest' and 'brightest', respectively). This pair is readily determined in
the course of a partitioning pass through the data. The darkest and brightest elements are especially informative with respect to images, but they are
not necessarily as effective with regard to obtaining compact constellations
in signal space. It is quite possible that the dark-bright pole constitutes a
relatively short diameter for a particular constellation. Likewise, the dark-
bright pole has somewhat constrained directional variation in signal space
over a series of subsets from successive partitions.

A modification can be made that makes the process of polar selection
substantially self-correcting with respect to compact partitions, irrespective
of the initial choice of polar pair. This modification retains the prior pole
for each successive subset, and pairs it with the most distant point in the
subset as its peripheral polar partner. Such polar pairs progressively pivot
directionally through signal space in an unrestrained manner, thus providing opportunity for and tendency toward realignment with larger diameters
of the constellations and thereby leading to more compact constellations
over a series of subdivisions. The β-scenario incorporates this modification.

A possible further consideration for the third stage of the 0-scenario is
that of using the coordinates of the mid-pole position in signal space as a
synthetic proxy for all the patterns in the nodal constellation. Despite the
common practice of using averages for aggregations, it could be argued
that an actual pattern should be preferred as a proxy over a synthetic one.
A practical possibility in this regard is to use the actual pattern points that
are closest to the prior quarter-pole positions. It may be noted that computation of mid-pole positions can be readily done retrospectively for comparative studies of the two methods.

The α-scenario has not had an overall measure for comparing complexity of constellations by which to track the trajectory of the third-stage splitting process through the successive cycles. A composite of the polar potentials of the nodal constellations can be formulated for this purpose. The utility for comparative purposes among images is improved by normalizing for both size of image and number of signal bands. Normalization for image size can be accomplished by using prominence of nodal patterns and normalization for signal bands can be done with number of bands as divisor. The composite complexity expression thus becomes Eq. 2.12,

$$Œ = (\sum p_k \Delta_k)/v \qquad\qquad (2.12)$$

where

p_k is the prominence of the kth third-stage pattern;

Δ_k is axial (Euclidean) distance between poles;

v is the number of signal band values.

Since splitting reduces axial distance and increases compactness, the complexity coefficient must decline stochastically as splitting progresses. In general, greater complexity implies larger average errors in approximating primitive patterns of the image by proxies of polypatterns. The discussion by Breiman et al. (1998) of nodal impurity measures for classification trees is relevant in this regard. It may also be noted that a single-pixel pattern would contribute nothing to this measure by virtue of having zero as the axial distance.

2.10 Tree Topology and Level Loss

Since the third-stage splitting process is influenced by both prominence and axial distance, it is may be of interest to examine the structural topologies of the partitioning trees created in that stage. Relevant relations of nodal constellations to complexity and sub-structure can be indicated as a topological triplet in Eq. 2.13,

$$constellations:complexity:connection \qquad\qquad (2.13)$$

where *constellations* is the number of nodal constellations as terminals of the tree, *complexity* is the complexity coefficient as given above but with the summation limited to the particular tree, and *connection* is the maximum number of intermediate nodes between the root and a terminal.

The members of the topological triplet are informative both individually and in combination. A large number of constellations indicates substantial subdivision. The number of constellations in relation to connections indicates the degree of consistency in subdivision of subdivisions. A 'bushy' tree will have the number of constellations large in relation to the number of connections, or conversely for a 'spindly' tree that is tall and thin. The complexity in relation to number of constellations indicates propensity for further subdivision under additional splitting cycles. Comparing the complexity across trees is indicative of disparity in pattern complexity.

Determining the variability of the B-level proxies about the A-level proxies is informative relative to level-loss of specificity associated with using the A-level independently of the B-level. A natural way of assessing this is by mean squared error of prediction incurred by using A-level proxies as estimates of B-level proxies. This is computed as the sum of squared distances between B-level proxies and A-level proxies divided by the total number of B-level patterns. It also provides an index regarding sensitivity of the landscape pattern to generalization. The evenness of this information loss can be assessed by computing mean squared error separately for each A-level pattern, and then computing mean, variance, and coefficient of variation for these error data.

2.11 γ-Scenario for Parallel Processing

The α and β scenarios of polypattern processes have been considered implicitly for sequential computing environments. The extraction of bi-level poly-patterns is an extended operation in such computing environments, with computation times that can run into hours for larger images on conventional desktop PC computers depending on the number of bands and B-level splitting cycles. On the other hand, subsequent analyses of polypatterns proceed much more rapidly than for conventional image data since many computations can be done in the domain of pattern tables instead of spatial lattices. Therefore, production operations involving numerous large images would require that pattern extraction times be reduced by an order of magnitude through parallel computing and related intensive computing facilities. In the γ-scenario, we consider prospects for pattern processes that are amenable to parallel computing.

The first stage of α and β scenarios presents barriers to parallel computing by the sequential nature. Therefore, further generalization is necessary. This generalization can be envisioned as 'potential pattern pools'. A potential pattern pool is configured to accommodate some maximum number Y of potential patterns which would be operationally dependent on the number of signal bands in a pattern, but should be several multiples of the 250 in the α and β scenarios. Such a pool would reside in a processor and would function in 'fill and filter' cycles in conjunction with similar pools in other processors.

The 'fill' operation would ingest pixels from the next available position of the image data file until the pool is at capacity. The pool would then be 'filtered' down to some designated fraction of its capacity by successively assimilating the pattern having weakest potential into its stronger neighbor. The space thus made available in the pool would then be refilled until the several processors had collectively exhausted the image data file. The processors would then proceed to collaborate in pooling their partial pools until a single pool having the desired number of potential patterns was reached.

Aside from parallel considerations, a modification of this nature is effective in countering the tendency of the β-scenario to produce several patterns having low prominence. A pool containing excess patterns is carried to the end of the first stage, and the top 250 potential patterns in order of prominence are extracted for the second stage. A pool having an excess of 10% has proven to be workable in this regard.

In the second stage, the processors would all work with the final set of potential patterns, but would divide the image file into sections for compilation and then cross-compilation of (proxy) pixel patterns. In both the first and second stages, it would be possible to have the processors work either asynchronously or synchronously.

In the third stage, a 'bifurcation brokerage' queue would reside somewhere in the system, and the processors would operate individual 'bifurcation buffers'. The bifurcation brokerage queue would prioritize the nodal constellations for splitting. An individual processor would have its bifurcation buffer filled from the front of the queue and proceed to split those particular patterns while purging them from the queue. Upon completing a set of bifurcations, the processor would adjust the common queue so that the subsets would take their respective places according to their eligibility in terms of the splitting criteria.

The fourth stage would complete the segmentation scenario for parallel processing by allocating part of the lattice to each processor for the necessary nested numbering of patterns. As for the β-scenario, redistribution of prominence over A-level patterns would be an optional aspect.

It might seem that greater computing power would also favor extension from bi-level polypatterns to tri-level polypatterns. However, adding additional tiers causes eighth-power exponential growth in the number of trees of patterns with considerable computational overhead in cross-indexing the levels. A third level is marginally feasible, but using a two-byte second level would be more practical. Even so, the additional byte at the second level would tend to be unevenly exploited.

2.12 Regional Restoration

Restoration is accomplished by placing the proxy signal vector for the B-level of polypatterns in each pixel. The result is a 'smoothed' or 'filtered' version of the multi-band data having somewhat less variability than the original, due to removal of the intra-pattern variability. This will usually have some beneficial effect of making the data less 'noisy'.

Fig. 2.4 shows a restored version of band 2 for the September 1991 Landsat MSS image appearing in Fig. 1.1. The quality of detail in such restorations relative to the resolution of the original provides support for our claim that B-level of polypatterns can model image intensities sufficiently well to constitute an image compression for purposes of landscape analysis. Resolution of the image in Fig. 2.4 is constrained by the large size of pixels in the parent image that are 60 meters on a side.

2.13 Relative Residuals

Residuals from restoration are of interest not only for purposes of spatial statistics, but also for determining whether there are portions of the image area that have been restored with less fidelity than others. For the latter purpose, it is desirable to have an integrated measure of residual that incorporates the effects of all signal bands. This can be accomplished by using the Euclidean distance between the proxy signal vector for the B-level pattern and the actual signal vector for the particular pixel. Fig. 2.5 shows this kind of multiband residual image for the restoration in Fig. 2.4.

The spatial pattern in the residuals is of particular interest. An 'ideal' pattern would be one of uniformity, indicating that the errors of approximation were evenly distributed over the image area. The next most favorable pattern would be a random 'speckle' distribution indicating that the errors of approximation constitute 'white' noise relative to environmental features and locations. Typically, however, there is some nonrandom spatial patterning of the residuals. Then it becomes necessary to refer to the image itself to determine what kinds of environmental features have the least fidelity in their representation. In the case of Fig. 2.5, the fringe areas of clouds are prominent with regard to residuals. Since clouds are typically nuisance features in the image anyhow, this is not a matter of concern. The stronger discrepancies are due to the spectral and spatial complexity of cloud fringes.

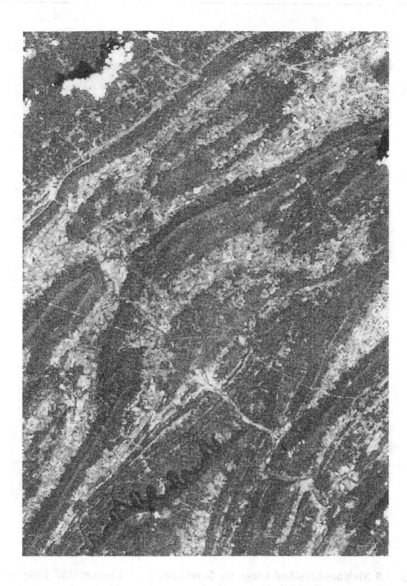

Fig. 2.4 Restored band 2 (red) of September 1991 Landsat MSS image of central Pennsylvania.

Fig. 2.5 Multiband residual image for September 1991 Landsat MSS image of
central Pennsylvania with darker tones indicating larger residuals.

It is also interesting to examine how the depth of pattern splitting from A-level to B-level varies across the image in relation to the relative differences in residuals. Fig. 2.6 shows this aspect of the pattern-based image modeling, whereby lighter areas have more splitting. It can be seen that the darker areas indicating larger residuals in Fig. 2.5 correspond with the lighter areas indicating greater depth of splitting in Fig. 2.6. Therefore, continuation of splitting cycles would also tend to focus on these patterns that have higher internal variability.

In his comparative work on clustering for data mining, Mirkin (2005) emphasizes recovery of original information from clusters as being the touchstone criterion for efficacy of clustering methodology. Our methods of pattern analysis fall generally under the statistical heading of clustering, although they involve non-conventional modalities and have some non-conventional goals even among the many versions of clustering.

Recovery is one of our goals, which we address in terms of restoration from compression as set forth above. We are in a position to conduct a thorough assessment of recovery in terms of residuals, with regard to both form of statistical distribution and spatial dispersion of residuals as demonstrated in this chapter. We are also in a position of advantage by having the capability to improve recovery through additional cycles of splitting if the residuals are deemed to be excessive.

In perhaps a somewhat counter-intuitive manner, we derive benefits from an assurance of less than complete recovery. One such benefit arises from avoiding exact electronic duplication of image data that may carry concerns for their proprietary nature. Our methodology produces image intensity models. The B-level patterns constitute a discretely valued model of the original image, and the A-level patterns provide a more generalized (less specific) model. Together, these models support multi-scale landscape analysis as well as selective enhancements for graphic emulations of pictorial images.

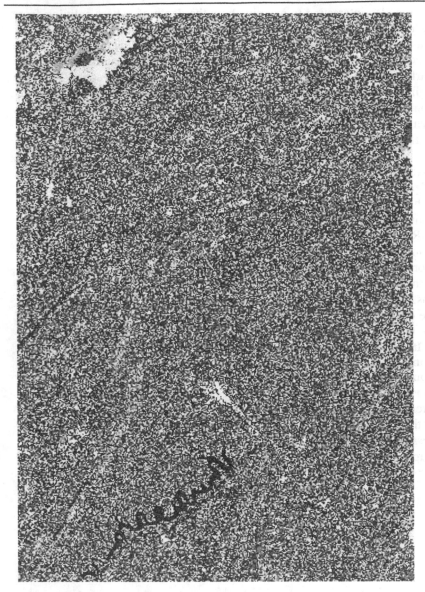

Fig. 2.6 Distribution of depth of splitting from A-level to B-level pertaining to residuals in Fig. 2.5 with lighter tones showing greater splitting.

2.14 Pictorial Presentation and Grouped Versus Global Enhancement

Direct display of A-level indexes as ordered overtones has been introduced earlier. Pictorial presentation of particular pattern properties involves triple indirection whereby look-up tables of pattern properties are used to prepare look-up tables of relative intensities that are used to prepare look-up tables that specify coloration for patterns. Since all of three of these tables are subject to change, there is a great deal of flexibility in the manner that patterns are portrayed. This makes possible enhancements of presentation graphics that cannot be obtained through conventional image analysis.

Enhancement is an image analysis term for operations that are applied to pixel properties in order to obtain pictorial presentations wherein certain aspects have more evident expression. In conventional image analysis, enhancement operations are applied globally to all pixels. Pattern presentations permit greater specificity whereby enhancement operations are applied selectively to a particular pattern or groupings of patterns, thereby avoiding alterations of the entire image. This even extends to generalizing the perception of patterns by displaying similar patterns the same to give them identical appearance. This is a considerable convenience for initial investigation of multi-scale characteristics of landscapes.

Appendix A suggests software that can be procured publicly without purchase for pictorial presentation of patterns. Since patterns are at least in part a perceptual pursuit, preliminary perusal of presentation protocols is prudent.

2.15 Practicalities of Pattern Packages

A prototype package for pattern processing is introduced in Appendix B. This is a modular package developed in generic C language to promote portability among platforms for computing. Specifications are submitted by editing a standard text file, which is then read by the respective module. The text file also contains imbedded instructions for making the appropriate substitutions. A Windows-style 'front-end' is also available as an alternative mode of managing most modules under Microsoft supported systems. The A-level of polypatterns is presented as byte-binary image information in a file having .BSQ extension, along with companion compilations of characteristics as textual tabulations in separate files. The B-level of polypattern information resides in a byte-binary file having .BIL extension, along with supporting tabulations that are also byte binary. Complete comprehension of the ensuing chapters may require reference to this software supplement, and even perhaps practice.

References

Breiman, L., J. Freidman, R. Olshen and C. Stone. 1998. Classification And Regression Trees (CART). Boca Raton, FL: Chapman & Hall/CRC. 358 p.

Burrough, P. and R. McDonnell. 1997. Principles of Geographical Information Systems. New York: Oxford. 333 p.

Chrisman, N. 2002. Exploring Geographic Information Systems. New York: John Wiley & Sons. 305 p.

DeMers, M. 2000. Fundamentals of GIS, 2nd ed. New York: John Wiley & Sons. 498 p.

Duda, R., P. Hart and D. Stork. 2001. Pattern Classification. New York: Wiley.

Forman, R. T. T. 1995. Land Mosaics: the Ecology of Landscapes and Regions. Cambridge, U.K.: Cambridge Univ. Press. 632 p.

Forman, R. T. T. and M. Godron. 1986. Landscape Ecology. New York: John Wiley & Sons. 619 p.

Fu, K. 1982. Syntactic Pattern Recognition and Applications. Englewood Cliffs, NJ: Prentice-Hall.

Gonzalez, R. and M. Thomason. 1978. Syntactic Pattern Recognition: an Introduction. Reading, MA: Addison-Wesley.

Gonzalez, R. and R. Woods. 2002. Digital Image Processing, 2nd ed. Upper Saddle River, NJ: Prentice-Hall. 793 p.

Jain, A., R. Duin and J. Mao. 2000. Statistical Pattern Recognition: a Review. *IEEE Trans. Pattern Anal. Machine Intell.*, vol. 22, no. 1, pp. 4-37.

Kelly, P. and J. White. 1993. Preprocessing Remotely-Sensed Data for Efficient Analysis and Classification. *Applications of Artificial Intelligence 1993: Knowledge-Based Systems in Aerospace and Industry, Proceedings SPIE 1993*, pp. 24-30.

Luger, G. 2002. Artificial Intelligence: Structures and Strategies for Complex Problem Solving. New York: Addison-Wesley. 856 p.

McGarigal, K. and B. Marks. 1995. FRAGSTATS: Spatial Pattern Analysis Program for Quantifying Landscape Structure. General Technical Report PNW 351, U.S. Forest Service, Pacific Northwest Research Station. 122 p.

Mirkin, B. 2005. Clustering for Data Mining. Boca Raton, FL: Chapman & Hall/CRC, Taylor & Francis. 266 p.

Myers, W., G. P. Patil and C. Taillie. 2001. Exploring Landscape Scaling Properties through Constrictive Analysis. J. Stat. Res. 35(1): 9–18.

Pao, Y. 1989. Adaptive Pattern Recognition and Neural Networks. Reading, MA: Addison-Wesley.

Pavlidis, T. 1977. Structural pattern recognition. New York: Springer-Verlag.

Podani, J. 2000. Introduction to the Exploration of Multivariate Biological Data. Backhuys Publishers, Leiden, Netherlands.

Schbenberger, O. and C. Gotway. 2005. Statistical Methods for Spatial Data Analysis. New York: Chapman & Hall/CRC. 488 p.

Simon, J. 1986. Patterns and Operators: the Foundations of Data Representations. New York: McGraw-Hill.

Tou, J. and R. Gonzalez. 1974. Pattern Recognition Principles. Reading, MA: Addison-Wesley.

Turner, M., R. Gardner and R. O'Neill. 2001. Landscape Ecology in Theory and Practice: Pattern and Process. New York: Springer-Verlag, Inc. 401 p.

Webb, A. 2002. Statistical Pattern Recognition. Chichester, England: Wiley.

3 Collective and Composite Contrast for Pattern Pictures

Contrast is one of the *4C* concerns for patterns in landscape images: *Contrast, Content, Context,* and *Change.* Contrast is important to pattern perception and distinguishing detail in an image. Contrast depends upon differences among image elements, but is not synonymous with inherent image information because it can serve to enhance expression of some elements at the expense of others. Thus, it may actually serve to suppress uninteresting information. Contrast can be controlled and is conveyed more effectively in color than in gray tones. Contrast enhancement is one of the ways that image analysts make images more interesting (Gonzalez & Woods, 2002), and is in some respects an almost artistic arena of image investigation. The pattern protocols that have been presented in previous chapters are amenable to *indirect imaging* which is conducive to contrast control.

3.1 Indirect Imaging by Tabular Transfer

The images used to illustrate previous chapters were developed by what may be described as distributive direct display using a *transfer function.* If the image data are represented as a horizontal axis and image intensity as a vertical axis, then data are distributively deflected to the display intensity axis by a transfer function as shown in Fig. 3.1. The transfer function can have any number of differently sloping sections, or be curvilinear, or even have portions that are horizontal. One that is linear in sections as shown is frequently said to be *piecewise linear.* The transfer function acts as a sort of filter.

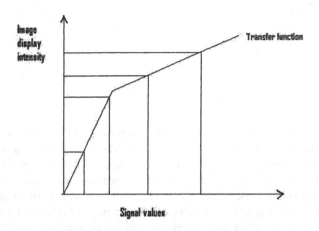

Fig. 3.1 Transfer function for determining display intensities of image data.

The steeper the slope of the transfer function, the greater the contrast for that subset of signal values. Conversely, gradual slopes give less contrast. A simple linear transfer function having 45° slope gives equal distribution of contrast over the range of signal values. Where the transfer function is horizontal, all signal values will have the same display intensity. Having horizontal sectors of the transfer function at the left end and the right end is called *saturation* at the dark end or the light end, respectively, because it homogenizes the extremes of intensity to permit more differentiation of detail in the mid-portion of the range of signal values. Using linear transfer functions is often called *contrast stretching* in the jargon of image analysis.

Since digital image data are usually recorded as discrete integers, a transfer function may actually be applied at the device level as a look-up table. The idea of *transfer table* can also be used more generally for indirect imaging that is filtered through a user-specified transfer table. Transfer tables offer a great deal of flexibility. Explicit transfer tables are used for *pseudo-color* displays of cellular maps to make them appear in the manner of images with a specified coloration of each numbered map element. This is the mode of imaging that we employ for our colorized A-level mappings of polypatterns. In so doing, we may either choose to have the numbered patterns portrayed as colors suggested by the image elements from which they were obtained, or we may choose to color code them in a more arbitrary *false-color* manner that suits a particular purpose.

3.2 Characteristics of Colors

Since pseudo-color imaging is central to controlling contrast in portraying A-level polypatterns as *pattern pictures*, we undertake a brief excursion into one scheme for characterizing color. Further detail and alternative schemes can be found in conventional digital image processing and remote sensing references (Cracknell & Hayes, 1991; Gibson & Power, 2000; Gonzales & Wood, 2002; Lillesand & Kiefer, 1999; Wilkie & Finn, 1996).

The scheme that we employ here is denoted as *HSI* for Hue, Saturation, and Intensity, although it is expedient to introduce these three in reverse order. As a prelude to this color characterization, it is essential to recall the tricolor synthesis of colors on RGB (Red, Green, Blue) display devices. Thus, a transfer table for casting an A-level lattice of patterns as an image will need to specify for each numbered pattern a triad of intensities for red, green, and blue on some color intensity scale. A common color intensity scale ranges over the integers from 0 to 255, but other ranges such as 0 to 1 can also be used for specification with the understanding that the tabular range is appropriately adjusted to accommodate a particular display device.

The sum of the three color intensity values for red, green, and blue will modulate the perceived brightness for a pixel without regard to the particular kind of color, and is therefore the *Intensity* part of the HSI characterization. Thus, increasing the intensity of any color component will increase the overall intensity. Intensity thus refers here to amount of light without regard to its quality.

It was also stated earlier that equal components of red, green, and blue give the visual sensation of a shade of gray. The minimum of the three intensities will thus blend with equal amounts of the other two in forming a gray component. This gray component that is $3 \times \min(R,G,B)$ tends to dilute the sense of color contributed by the other two, making it impure or *unsaturated* in a pastel sense. The S in HSI stands for *saturation*, with saturation being greatest when only two of the three colors are active so that there is no grayness to the mix.

The remaining mix of two unbalanced color components is responsible for the kind of color or *hue* that is perceived, with hue being the H in HSI. An equal mix of red and green intensities in the absence of any blue will appear yellow, with the brightness of the yellow being determined by the total of the intensities for red and green. An equal mix of red and blue in the absence of green will appear magenta. An equal mix of blue and green in the absence of red will appear cyan. In this additive color mode, yellow is said to be the complement of blue, magenta to be the complement of green, and cyan to be the complement of red. Unequal mixtures of two colors will give some other hue.

3.3 Collective Contrast

We come now to the respect in which contrast of polypatterns is different from that of the parent image of primitive pixels. All pixels that might have had unique signal vectors have been generalized by the pattern processing so that their (proxy) property point is shared with other pixels in the pattern. All non-spatial aspects of contrast contained in the patterns can be captured in a table having as many rows as there are patterns. With the PSIMAPP software introduced in Appendix B, the first (index) column for the table has pattern number, the second column has pixel population (frequency) for the pattern, and the remaining columns have properties of the proxy for the pattern. It is natural to have the order of rows be that of pattern number. Such a table for the A-level of polypatterns has 250 rows, and carries a .CTR extension on the file name. Likewise, non-spatial contrast characteristics of the patterns with regard to signal values can be computed from the table without recourse to the lattice of positional mappings for the patterns. These collective computations are much faster and easier than doing the corresponding computations from an original multi-band image file. Given the tabular nature of the contrast information, indirect imaging with transfer tables has added advantages.

The first aspect of this advantage lies in ability to adjust the contrast relations among the patterns algorithmically in an iterative manner using statistical criteria of contrast that would be essentially impractical for full images of pixels. Since this mode of contrast control is not conventional for image analysis, we give it special terminology as *auto-adaptive contrast control*.

In working with contrast of polypatterns through PSIMAPP, the first step is to prepare a pair of tables specifying the relative brightness among the A-level patterns (see also Appendix B.3). The brightness values are scaled from zero to one, which confers some computational advantages. The first of these two tables has relative brightness of the respective signal bands as columns, and has a .BRS extension on the file name.

The second table has retrospective mathematical formulations that amalgamate the signal bands to obtain *integrative image indicators* of environmentally interesting contrasts. This table has a .BRI extension on the file name. Both of these tables have a fixed-width textual form that is easily imported into spreadsheets.

Auto-adaptive contrast control is optional in preparing the .BRS and .BRI relative brightness files. The version of auto-adaptive contrast control incorporated in PSIMAPP software is a statistically guided sequential linear stretch with saturation on the high side of the brightness scale. Saturation essentially entails truncating the signal range for proportional transfer of intensities to display devices. Statistical character of contrast provides the point of departure. Contrast is perceived as differences in an image. Differences are due to variation, and variation is gauged statistically as *variance*. Increasing contrast is thus associated with increasing variance of intensities in the image. Since maximum intensity is accorded to patterns having signal values above the truncation threshold, saturation induces areas of uniformity in the displayed image. When areas of uniformity become large, then variation in the image is reduced. There is also a concern for tonal balance whereby the average intensity should not be extreme. Proportional scaling after truncating the range at lower values has the effect of increasing the average intensity over the image.

The algorithm for auto-adaptive contrast control finds the two highest levels of intensity for patterns at a current stage, and truncates the range at the lesser one so that the greater one becomes saturated. The overall mean and variance under this modification are computed. The modification is retained if it increases the overall variance and keeps the average intensity level for the image at or below 0.4 as mean value. If the modification is retained, the two highest remaining levels of intensity for patterns are found and the process repeated. If a modification is not retained, then the process terminates. Images having particular areas that are especially bright (such as the tops of clouds) tend to benefit most from this auto-adaptive contrast control.

3.4 Integrative Image Indicators

The .BRI brightness file contains a suite of eight integrative image indicators, some of which pertain primarily to spectral signals of particular wavelengths. The first of these integrative indicators is a unit scaling of A-level pattern numbers for the ordered overtones. This corresponds to direct display of A-level polypatterns as illustrated in Figs. A.10 and A.11 of Appendix A.

The second integrative indicator is a reverse scaling of the first one. This reverses darkness and brightness in the image.

The third integrative indicator totals intensity across all signal bands of an A-level pattern. Whereas the first indicator is based on order of aggregate intensities, this third indicator also takes into account the strengths of the intensities as well as their order.

The fourth and fifth integrative indicators are optional, and their meaning depends upon the nature of the signals in the dataset. Each of these gives total intensity for a subset of the bands. If the dataset contains spectral data, it is normally intended that the fourth be a total for the visible bands and the fifth be a total for the infrared bands.

The sixth, seventh and eighth synthetic signals have meaning primarily in relation to remote sensing of vegetation. The sixth is a modified pattern version of the so-called NDVI (Normalized Difference Vegetation Index) for which the computing formula is (Infrared-Red)/(Infrared+Red). This ratio tends to be high (bright) for vegetated areas and dark for areas lacking vegetation, with broadleaf foliage expressing somewhat more strongly than needle-leafed foliage. For present purposes, experience has shown an improved gradation if NDVI is multiplied by its square root. Accordingly, this modification has been adopted. This index is illustrated in Fig. 3.2 for a September 6, 1972 Landsat MSS of central Pennsylvania. At that time of the early fall season, the trees on the ridges still retained their foliage whereas crops in the agricultural valleys had been harvested or were senescent.

The seventh is a combination of total brightness, infrared subtotal brightness, and modified NDVI in a formulation that is intended to lend some emphasis to forest comprised of needle-leafed trees. This 'conifer' index is a novel conception in the current work. It is computed by Eq. 3.1,

$$\text{Total brightness} + (0.5*\text{NDVI}) \times (1.0-\text{IR})^2 \tag{3.1}$$

with truncation to a maximum value of one. This indicator is shown in Fig. 3.3 for the same situation as the previous one.

The eighth is an inversion indicator that tends to make dark areas bright, particularly for water. It is formulated (with modified NDVI) as Eq. 3.2.

$$1 - \max(\text{total brightness, visible brightness, IR brightness, NDVI}) \tag{3.2}$$

This has an intermediate value unless the maximum of the other values is extreme. This indicator is illustrated in Fig. 3.4 for the same situation as previous ones. Note that cloud shadows have a similar appearance to water surfaces as seen through this indicator.

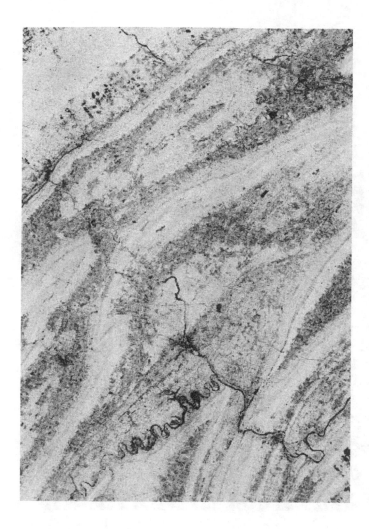

Fig. 3.2 Modified NDVI indicator for portion of a September 6, 1972 Landsat MSS scene in central Pennsylvania.

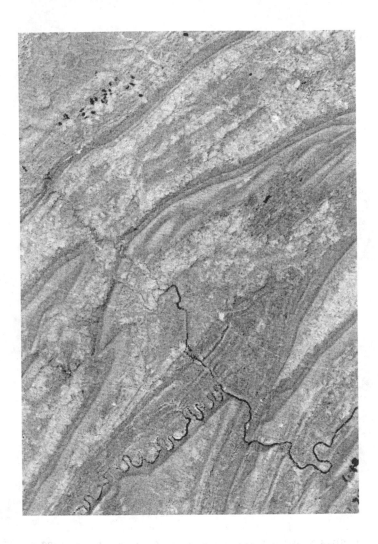

Fig. 3.3 Conifer enhancing indicator for portion of a September 6, 1972 Landsat MSS scene in central Pennsylvania. Conifers are localized and not obvious in this area.

Fig. 3.4 Inversion indicator for portion of a September 6, 1972 Landsat MSS scene in central Pennsylvania with water and cloud shadows appearing bright.

3.5 Composite Contrast for Pattern Pictures

As stated earlier, contrast is most effectively conveyed in color. Three different bands or indicators can be assigned to the respective components of an RGB pseudo-color image using polypattern software facilities (see also Appendix B.4). Considerable creativity can be exercised in this regard, but contrast in the color image is increased when the (gray-tone) appearances of the three selections are substantially different. If the corresponding signal bands are available, an obvious choice would be to mimic the preference of nature by using blue signal for blue, green signal for green and red signal for red to give the *true color* or *conventional color* that we are accustomed to perceiving with our eyes. One drawback in this regard is that blue wavelength has high atmospheric interference (which makes the sky blue) so that images made from space tend to be quite hazy as seen in a blue band.

A tri-color composite that is well known to image analysts avoids this difficulty with blue haze while also taking advantage of the high infrared reflectivity in healthy broadleaf plants. This composite places green signal on blue, red signal on green and a near-infrared signal on red. The short designation for such a composite is CIR for Color Infra-Red. Since healthy broadleaf plants absorb much of the red solar radiation for photosynthesis, their coloration becomes a blend of red and blue to form magenta. This reddish tone of vegetation is distinctive in such an image. This kind of image is sometimes also called a *camouflage detection* image because the magenta tone fades rapidly to a whitish tone when the plants are placed under heavy stress. Thus, cut vegetation used to conceal something by covering will lack the characteristic magenta tone.

Even with only four signal bands and eight indicators, there are a large number of possible combinations that could form the tricolor composite. Algorithmic assistance in choosing triplets that complement each other is thus in order. A strategy developed in conjunction with the polypattern approach (see also Appendix B.5) is to choose first the band or indicator having highest variance (weighted by prominence). The next band chosen is the one having greatest sum of squared Euclidean (weighted) distances from the first band, but avoiding anything that exceeds 0.9 with regard to unsigned correlation coefficient because of redundancy. The third band is similarly selected according to summations of squared distances from both of the first two bands. Next best replacements for the selections are also considered as alternatives. This algorithm does not suggest which of the selections should be assigned to a particular color in forming the actual composite image. Focus is only on high-contrast combinations, not how they are combined. The choice of color assignments is left as the artistic purview of the analyst. For example, running the algorithm for the image data used in Figs. 3.2-4 gives a primary recommendation for the combination of red signal, vegetation indicator and inversion indicator. Assigning the red signal to red, the vegetation indicator to green and the inversion indicator to blue does indeed give a strongly (perhaps even glaringly) contrasting image. Forest appears bright green. Agriculture appears bright orange. Urban centers appear reddish orange. Water and cloud shadows appear blue.

3.6 Tailored Transfer Tables

As indicated at the end of Chapter 2, pattern-based models of images have a fundamental advantage over the initial images with respect to contrast control of components. Even though the results differ for different types of patterns, all of the operations considered thus far in this chapter have been applied globally to all patterns. In transitioning from the modeling methods of the previous chapter to the tonal tabulations of this one, it was noted that there are three types of look-up tables involved. The first type of table (with .CTR extension) provides properties of patterns. The second type of tables (.BRS and .BRI) treats the properties as intensities of image indicators for relative ranges. The third type of table (.CLR) translates indicator intensities into tricolor tones.

The PSIMAPP software has been configured so that all three types of tables have a simple textual structure that can be directly altered with a text editor. Having gained some familiarity with the layout of these tables, there is the capability for selectively specifying the way in which a particular pattern or group of patterns is to be portrayed. The patterns of interest can be located either by examining the tables, or by using a display facility that allows one to query the pattern number at a pixel position. For example, Fig. 3.5 shows the bioband patterns accounting for the top 5% of overall richness in white (additional area is nearly white).

Selecting patterns for distinctive display according to special properties makes an excellent topic of transition into *content* which is the subject matter of the next chapter. The perceived sense of differences in an image is also influenced by the fragmentation, juxtaposition and interspersion of the contrasting pattern elements. These spatial aspects, however, are considered subsequently under the topic of *context*.

Fig. 3.5 Bioband patterns in white accounting for top 5% of cells with respect to overall species richness. Additional area is nearly white, which could be clarified by using color.

References

Cracknell, A. and L. Hayes. 1991. Introduction to Remote Sensing. New York: Taylor and Francis. 303 p.

Gibson, P. and C. Power. 2000. Introductory Remote Sensing: Principles and Practices. New York: Taylor and Francis. 208 p.

Gonzalez, R. and R. Woods. 2002. Digital Image Processing, 2nd ed. Upper Saddle River, NJ: Prentice-Hall. 793 p.

Lillesand, T. and R. Kiefer. 1999. Remote Sensing and Iimage Interpretation, 3rd ed. New York: Wiley. 750 p.

Wilkie, D. S. and J. T. Finn. 1996. Remote Sensing Imagery for Natural Resources Monitoring: A Guide for First-Time Users. New York: Columbia University Press. 295 p.

4 Content Classification and Thematic Transforms

A question that arises naturally when examining landscape patterns concerns the sorts of landscape elements that comprise the several patterns. Since this is also an ecologically and economically important question, *content* is the next of the *4C* aspects (*contrast, content, context* and *change*) to be considered. Whereas we have been able in much of the foregoing to largely relegate software matters to the Appendix, it is less effective to do so in regard to analysis of content. The focus in this chapter with regard to content is on designating different portions of landscapes as belonging to one or another of a mutually exclusive set of categories or classes with respect to composition. This is often encompassed in the term *thematic* mapping.

It is acknowledged at the outset that the ideal of 100% correct classification is virtually never achieved. The enterprise of content classification thus becomes one of balancing errors of *omission* and *commission*. Omission error is when an occurrence goes unrecognized. Commission error is when an occurrence of one kind is wrongly designated as being of another kind. Accuracy can also be reported differently from map user versus map producer perspectives. The user perspective concerns the percentage of class occurrences shown on a map that are correct. The producer perspective concerns the percentage of occurrences for a class on the landscape that are shown correctly on the map. A map could be 100% accurate for a particular class in the producer sense by virtue of showing all of those occurrences as the correct class, but still be inaccurate in the user sense due to commissions of also showing incorrect inclusions of other kinds as belonging to the class (Congalton and Green, 1999).

Consistency of occurrence and appearance for landscape patterns is fundamental to compositional analysis of landscape content. The idea of consistency for landscapes rests on the assumption that a particular phenomenon will express in a specific manner as long as similar circumstances pertain, but that such expression also entails some intrinsic variability. This consistency applies to characteristics of expression in spectral reflection as well to context. Since spectral reflection is a function of illumination, different conditions of illumination constitute different circumstances for expression of landscape phenomena as patterns in images. For example, forest on the sunlit side of a ridge appears differently than forest on the shaded side. In remote sensing, the idea of consistency in appearance is so fundamental that spectral expression is spoken of as a *signature* for the corresponding phenomenon.

Although contextual aspects of spatial structure in patterns constitute the topic of another chapter, some of the implications must also be taken into account for classifying content. Patches, corridors, gradients, disturbances, and progressions of succession are characteristic of landscape spatial patterns (Forman, 1995; Forman and Godron, 1986; McGarigal and Marks, 1995; Turner et al., 2001; Wilson & Gallant, 2000). Lines of demarcation for spatial transitions of patterns are sometimes abrupt, but more often gradational. Further complications arise when a single pixel spans transitions of content, thus giving rise to so-called 'mixed pixel' effects. Computers excel in capability for distinguishing among spectral expressions, but human vision still holds an advantage with regard to detecting differences in spatial aspects of patterns. The latter advantage does not extend, however, to objective description of the differences in patterns that are noticed spatially. Therefore, it becomes advantageous to form a partnership between human analyst and computer whereby each contributes most in the area of advantage. Since most methods of computer-based classification for images also rely on some initial human designations to serve as *training sets*, it is appropriate to focus first on the human interpretive aspect.

4.1 Interpretive Identification

There has been a sort of continuing tension between two modes of working with image information. The mode that is usually called photo-interpretation has a longer tradition of mapping and relies primarily on the capabilities of trained human vision to recognize environmental features on the basis of image clues such as size, shape, tone, texture, shadow, pattern, and location/association. The other mode relies minimally on human visual interpretation and more on ground reference information (or so-called ground truth) as a point of departure for quantitatively comparative computer vision wherein pixel information is matched to the statistical properties of known samples.

Human photo-interpretation originally used image documents such as paper prints and film transparencies whereby the analyst would delineate features of interest on clear overlay material with a fine-line marker. A more technologically sophisticated version of this is to substitute a computer display of a scanned image document or a multiband image file instead of the hardcopy document, and a mouse-controlled cursor instead of a marker. This kind of interactive content delineation requires fairly sophisticated display software like that found in commercial GIS packages. A well-conceived color composite pattern picture display of the A-level from polypatterns as discussed in Chapter 3 is advantageous in this regard. There is less computational overhead of data to be processed as the interactive work takes place than there would be with actual multiband image data, which makes the procedure more rapid. Some software systems that lack full image handling capability can handle the simpler pseudo-color form of the pattern picture. Even if full image handling capability is available, the pattern picture has prepackaged color so that the mapping analyst need not be concerned with choosing a color scheme.

An A-level pattern picture enables another mode of interactive mapping that is not available with actual multiband image data. This also requires sophisticated software for map displays like those of ESRI's ArcGIS facility. The interpretive mapping scenario entails overlays in a viewer. The A-level (pseudo-color) pattern picture is brought into the viewer as a base layer. A second copy of the pattern picture is then placed on top of the first, and made entirely transparent so the user sees the base layer. A cursor query facility is used to determine what A-level pattern number resides at a location of interest. Using the second copy of the pattern picture, the analyst can examine the spatial structure of the pattern and its relation to other patterns. This combination of information is often sufficient to assign the pattern to one of the legend categories for which content mapping is underway. The top layer can be turned on and off as needed so that the developing thematic map of legend colors can be checked in relation to the pattern picture comprising the bottom layer.

Fig. 4.1 illustrates a map that was developed in this manner using the 'bioband' dataset of multiple indicators on biodiversity in Pennsylvania that was introduced in Chapter 1. The intent for mapping was to determine the occurrence of chronic degradation in upland and lowland habitats of Pennsylvania. The distinctions are made on the basis of richness of habitats for species in six taxonomic groups: mammals, birds, amphibians, snakes/lizards, turtles, and fishes. Lowlands are recognized as having strong representation of aquatically associated species. This gives rise to four categories considered as: (1) viable lowland habitats, (2) viable upland habitats, (3) degraded lowland habitats, and (4) biotically impoverished. The general paucity of biota does not provide a basis for distinguishing upland from lowland in the fourth category. The third category is

Fig. 4.1 Generalized map of habitat degradation in Pennsylvania from richness of habitats for vertebrate species. Light gray = viable lowland habitats; medium gray = viable upland habitats; dark gray = degraded lowland habitats; black = biotically impoverished.

distinguished by generalist species in the uplands with some remnants of distinctive lowland species.

A point of beginning for this purpose was to examine the .CTR table of central values for A-level patterns to note which pattern numbers had notably high or low central values for the different types of biota. This can be done conveniently by importing the textual .CTR file into a spreadsheet. The table has A-level pattern number in the first column, pixel population for the pattern in the second column, and each subsequent column gives a typical value for the respective signal band.

The tabular information was supported by experimentation with rendering of pattern pictures that tended to contrast upland groups and lowland groups. An 'upland indicator' was created analogous to the NDVI vegetation indicator by treating mammal richness as if it were the infrared signal band and fish richness as if it were the red signal band. This tended to brighten the uplands while darkening the lowlands. A particularly informative rendering was obtained by making this upland indicator appear as red, with slightly subdued amphibian richness being green and strengthened turtle richness being blue. A rendering that helped to differentiate various lowland settings was to represent amphibian richness as red, turtle richness as green and fish richness as blue. The more obvious situations were given category color codes to start the mapping process. Unmapped locations in these vicinities were then identified with respect to A-level pattern number and their spatial arrangements studied to make categorical assignments. The remainder of the area was thereby mapped in progressive manner. A GIS analysis environment thus provides excellent opportunities for interacting with pattern pictures for mapping purposes.

4.2 Thematic Transforms

Fig. 4.1 provides a basis for introducing a special display capability as a variation of pattern pictures through *thematic transforms* by doubly indirect imaging. The mechanism for implementing this is incorporated as a special feature of the software module for which an overview is given in Appendix B.4.

The *thematic transform* is a vector (list) of thematic category assignments for A-level patterns. In the implementation that is covered in the appendix, this list takes the form of a simple textual file having a .THM extension as the end of its file name. There is a line in this file for each of the 250 sequentially numbered A-level patterns, and an additional initial line for zero as a flag for missing data. Category numbers serve as entries in this thematic transform file. The thematic category to which an A-level pattern has been assigned is placed as the only entry on each line. A zero (0) is entered on the line if a pattern has not been assigned to any of the thematic classes. Because of the initial line for missing data, the class (category) code number corresponding to the ith pattern is placed on line $i+1$ of the thematic transform file.

The thematic transform constitutes a first stage of indirect imaging. A second stage of indirection assigns a color to each thematic category through a *thematic tonal transform table* contained in a textual file having .TON extension on its file name.

There is one line in the thematic tonal transform table for each thematic category, with entries on the line being delimited by spaces (not tabs). The first entry on a line is the thematic category number. The second is a decimal number between 0.0 and 1.0 inclusive that indicates the relative 'dose' of red. The third indicates the relative component of green, and the fourth indicates the component of blue. It is possible to remove a category from the map temporarily by putting a negative sign on the category number for that line. Since gray tones are used in Fig. 4.1, the colors have equal contributions to the tone. The .TON table for this example is shown in Table 4.1.

Although the software module is the same as introduced in Appendix B.4, checking the *View thematic categories* option causes a reconfiguration of the initial dialog box as shown in Fig. 4.2 so that the thematic transform and thematic tonal transform files can be selected. Any patterns not assigned to a thematic category are pictured using the gray-scale selected near the bottom of the dialog box.

Table 4.1 Thematic tonal transform table used for producing Fig. 4.1.

```
1   0.0   0.0   0.0
2   0.3   0.3   0.3
3   0.6   0.6   0.6
4   0.9   0.9   0.9
```

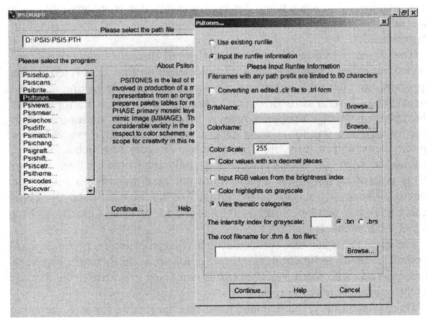

Fig. 4.2 Thematic version of dialog box replacing Appendix B.4 standard mode.

4.3 Algorithmic Assignments

It is often seen as being desirable to expedite the mapping process by increasing the degree of automation. The conventional 'supervised' scenario for computer-assisted mapping from multiband data has a skilled image analyst acquire so-called 'ground truth' information from maps, documents, and/or field work by which to locate samples of each category in an image display to serve as 'training sets' from which to extract statistical characterizations of signal patterns that serve as 'signatures' for quantitative pattern matching to assign a category classification for the other pixels comprising the image. The quantitative method or rule set for assigning pixels to categories can range from very simple to very sophisticated (James, 1985;Tso & Mather, 2001).

At the most simple extreme is a set of thresholds or cutoffs specifying the lowest and highest values to be allowed for the class in each band. This simple method is often given the imposing name of parallelepiped, which is basically just a virtual 'box' in the multi-dimensional signal space where bands serve as axes. The criteria for choosing the thresholds on the bands from training set data can be equally varied from statistical to empirically judgmental. These may use statistical measures and theoretical distributions or visual perusal of histograms.

A step up in sophistication is to compute the Euclidean distance of each candidate pixel from the central value(s) for the training set(s) of the categories, and assign the category for which this distance is minimum. Setting a specific threshold distance for each class gives rise to virtual 'bubbles' instead of boxes. Several variations on the distance idea are equivalent to rotating and rescaling the signal axes relative to the original bands so that the ratio of between-category to within-category variation is increased before computing distances in the transformed 'space' perspective. The Bayesian 'maximum likelihood' version uses a typical statistical assumption that the variation within categories conforms to an idealized multivariate 'normal' distribution. Even this maximum likelihood strategy, however, can also be conceived as a kind of generalized distance involving relatively complicated transformations of signals.

Even empiricism is not necessarily simple. Neural network approaches drawn from the field of artificial intelligence are computationally sophisticated but essentially empirical. While it is not the purpose here to undertake an exhaustive coverage of classification algorithms, it is important to note that the more sophisticated ways tend to place greater reliance on larger and presumably representative selections of training sets. Also, the analyst often has only limited capability to override the automatic assignments.

These conventional approaches become available at the B-level of polypatterns through various existing software packages when the restoration capability is used for generating 'smoothed' approximations to original bands. In so doing, one should not lose sight of the fact that 'smoothing' works to reduce the within-category variation among pixels since all pixels in any given B-level pattern will be identical. Likewise, the entire pixel population of a B-level pattern will necessarily be assigned to the same class. The focus here, however, is on a strategy that is geared to the A-level patterns and places a premium on having the analyst retain final control of category assignments while having the computer make recommendations in this regard.

4.4 Adaptive Assignment Advisor

A-level patterns are polypatterns comprising constellations of B-level patterns. The pattern parsing process allows some A-level patterns to enfold many B-level patterns, while others encompass only a few or even just one B-level pattern. Furthermore, the A-level patterns themselves typically have substantially different sizes of their pixel populations or prominence. The second column of the .CTR central values table shows the sizes of the A-level pixel populations, and the second part of the .PSI file shows the number of B-level patterns in each A-level pattern. In preparing to undertake categorical classification of content in terms of polypatterns, it is advisable to make note of these size differences and exercise additional caution in assigning the more prominent ones that will account for substantial sections of the image area.

The first concern in categorizing polypatterns is to capture characteristics of complexity from the parent pixel primitives. This is accomplished by tabulating supplementary statistics as described in Appendix B.2. There are two tables of such statistics that summarize how pixel property points are scattered in signal space. A suite of statistics pertaining to variability among constellations comprising patterns is recorded in a textual table as a .VAC file. There is a line for each of the A-level clusters with nine items of information (fields) on each line. The first field is the A-level pattern number. The second field tells number of pixels comprising the pattern. The third field is the minimum standard deviation among the signal variables (bands) for the pixels in the pattern. The fourth field is the maximum standard deviation among signal variables for pixels in the pattern. The fifth field gives the maximum (Euclidean) distance from a pixel in the pattern to the central position (centroid) of the pattern, which can be considered as a sort of radius in signal space. The sixth field is minimum distance from the pattern centroid to a pixel that is not in the pattern. The seventh field is the number of the pattern to which the closest pixel in the sixth field belongs. The eighth field is the minimum distance from the pattern centroid to the centroid of another pattern. The ninth field is the number of the pattern having the closest centroid-to-centroid distance. All of these are based on Euclidean distance. This .VAC table contains considerable information on the compactness and separability of patterns, not all of which is necessarily used in classification. The primary items for classification are the minimum and maximum standard deviations of signal (band) components for the pattern.

The other textual table of summary statistics is in a file with .MEN extension on the name. This second file is like the .CTR in showing central positions of the respective patterns. The difference is that these central positions are computed entirely as arithmetic means of pixel positions. Therefore, the .MEN file has signal centroid positions in a proper sense. The original .CTR file is better said to contain 'central values' as a somewhat nonspecific terminology. If desired, the information in the original .CTR file can be archived under a different name and then replaced by a copy of the .MEN file with the name changed to have the required .CTR extension. It should be routine procedure to obtain these two files as part of polypattern preparation so that they will be available if and when needed for classificatory purposes. If this practice is followed, the polypatterns become largely self-contained so that usual usage no longer depends on possession of the original multiband image data.

PSIMAPP software for polypatterns includes an adaptive assignment advisor (PSITHEME) for computer-assisted thematic classification that is intended for partnership with a human analyst. The analyst uses a viewer such as MultiSpec (Appendix A) having an image query capability (see MultiSpec *View* menu for *Coordinates View* and *Processor* menu for *List Data*) to determine the pattern number that occupies a location for which the correct thematic category is known from ancillary information such as maps or 'ground truth'. The pattern becomes a prototype or 'training set' for a category to be considered in conjunction with information on interpattern variability from the tables described in the preceding paragraphs.

In this synergistic scenario, the determinations for propensities of patterns relative to thematic category classes are of four qualitative kinds, with two kinds being made by the analyst and two by the program. The two requiring analyst action are *definite* and *reject*. *Definite* is an explicit declaration by the analyst that the pattern is a member of the class and thus constitutes a class prototype or training set. *Reject* is a counterpart declaration by the analyst that the pattern *does not* belong to the class in question and is not eligible for consideration as a candidate for membership in the class. A *probable* designation for a pattern relative to a class comes from the advisory algorithm and is strongly supportive of membership to the degree that the probable patterns are adaptively incorporated into pattern prospecting for additional class candidates. A *possible* designation for a pattern relative to a class comes from the advisory algorithm but is tentative to the degree that possible patterns do not play a role in pattern prospecting for additional class candidates. A pattern can be designated by the analyst as a *reject* for any number of classes, but can be designated for membership in only one class and in only one way with *definite, probable* and *possible* being the mutually exclusive ways of designation. The reject designation, however, has two special variations. A negative *reject* designation prevents a pattern from entering a class on a *probable* basis, but permits entry on a *possible* basis. A *reject* designation of 0 (number zero) limits that entire class to having only *definite* pattern members, with the consequence that specification of membership for that class is done entirely by the analyst without assistance from the algorithmic advisor.

Probable determinations are made on the basis of minimum and maximum standard deviations of signal variables (bands) for the patterns from the .VAC file. *Possible* determinations are made solely on the basis of maximum standard deviations for the patterns, also from the .VAC file. All *probable* assignments are made before any *possible* assignments. For each class the advisory algorithm locates the unassigned pattern centroid that is closest by Euclidean distance to a member of the class and is not a *reject* for the class. Classes are then considered for augmentation in order of proximity for the closest unassigned pattern.

A member is added as *probable* by either single or multiple modes of linkage. For single linkage mode, the current and candidate couplings are assessed by summing their minimum standard deviations and then multiplying the sum by a user-specified *alpha* scale parameter to obtain a threshold distance. If the centroid of the candidate is within the threshold distance of the centroid for the current member, then the candidate follows the current member into the class. For the mode of multiple linkages, the maximum standard deviations are used instead of minimum and the candidate must be within threshold distance of at least two current class members.

Single linkage is used for *possible* assignments, with the sum of maximum standard deviations being multiplied by a *beta* scale parameter to obtain the threshold distance. Both the *alpha* and *beta* scale parameters are controlled by the analyst, with a value of 1.0 being the default for the dialog box as shown in Fig. 4.3. There is a further option to make the beta scale parameter negative, which causes *possible* memberships to be omitted from the thematic transform table, and thus also from the resulting pseudo-color map.

The analyst employs a text editing facility to prepare a file of training information containing the *definite* and *reject* (if any) specifications. This file containing the training information has a simple text form that does not permit presence of tabs or formatting characters for word processing. There is a section for each class, which is introduced by a line containing the word CLASS followed after one or more spaces by the class number. The next line has the word DEFINITE, and the subsequent line(s) hold(s) the space-delimited list of *definite* patterns for the class. Any *reject* patterns for the class would be specified in similar fashion by a leading line containing only the word REJECT and subsequent line(s) listing the respective patterns that are to be excluded from the class. Any line in such a file that starts with an exclamation mark is treated as a comment and otherwise ignored. The name of the training file is to be chosen by the analyst and specified as the Train File in the dialog box.

The advisory module takes the file of training information as input along with files of pattern statistics. The module revises the training file to incorporate updated membership status as one of its outputs, including a listing of patterns that lack any current membership along with information on alteration of parameters that might expand memberships. After viewing and mapping the suggested memberships, the analyst can edit the *definite* and *reject* declarations and then make another run with alterations of *alpha* and *beta* scale parameters if appropriate. Information on *probable* and *possible* declarations from any previous run need not be removed since it is simply ignored in the next run and revised accordingly. A thematic transform file having a .THM ending is also produced for mapping purposes as described earlier. The thematic transform file is referred to as a Class Index File in the dialog box.

Mapping of water surfaces in the September 1991 Landsat MSS subscene for central Pennsylvania can serve to illustrate the process. Table 4.2 shows the training file for this purpose, with MultiSpec having been used to select clusters 6 and 7 from a large reservoir called Raystown Lake.

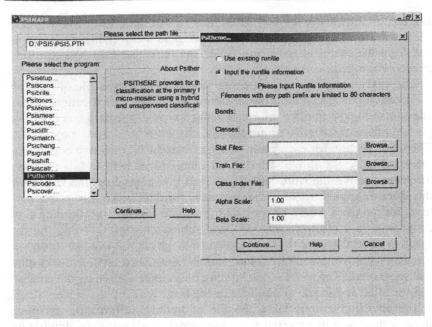

Fig. 4.3 Main dialog box for assignment advisory module.

Table 4.2 Training file for computer-assisted mapping of water surfaces in September 1991 Landsat MSS sub-scene of central Pennsylvania.

CLASS 1
Definite
  6 7
Probable
  8 3 10 11 9
Possible
  5
!
! Unclassified
! 1   2   4 12 13 14 15 16 17 18 19 20 21 22 23
! 24 25 26 27 28 29 30 31 32 33 34 35 36 37 38
! 39 40 41 42 43 44 45 46 47 48 49 50 51 52 53
! 54 55 56 57 58 59 60 61 62 63 64 65 66 67 68
! 69 70 71 72 73 74 75 76 77 78 79 80 81 82 83
! 84 85 86 87 88 89 90 91 92 93 94 95 96 97 98
! 99 100 101 102 103 104 105 106 107 108 109 110 111 112 113
! 114 115 116 117 118 119 120 121 122 123 124 125 126 127 128
! 129 130 131 132 133 134 135 136 137 138 139 140 141 142 143
! 144 145 146 147 148 149 150 151 152 153 154 155 156 157 158

```
! 159 160 161 162 163 164 165 166 167 168 169 170 171 172 173
! 174 175 176 177 178 179 180 181 182 183 184 185 186 187 188
! 189 190 191 192 193 194 195 196 197 198 199 200 201 202 203
! 204 205 206 207 208 209 210 211 212 213 214 215 216 217 218
! 219 220 221 222 223 224 225 226 227 228 229 230 231 232 233
! 234 235 236 237 238 239 240 241 242 243 244 245 246 247 248
! 249 250
!
!Forward factor ranges after Implicit stage:
! Range for class 1 = 0.938230 - 2.119478
```

A simple thematic tonal transfer table as a .TON file consisting of the single line: 1.0 1.0 1.0 was then prepared to render water surfaces as being white on a gray-tone image background. The .TON file must have the same base name as the .THM file as specified in the dialog box. As explained earlier, a thematic tonal transfer table has a line for each class. The first entry on the line is class number. The second entry is a relative red component ranging from 0.0 to 1.0, the third is a relative green component, and the fourth is a relative blue component. In this case, the equal and maximum components cause a rendering of the water category as white in the resulting depiction of Fig. 4.4. A blue color would be more natural and would also avoid possible confusion of water with clouds; but the distinctive cloud shapes make this less of an issue for illustrative purposes without incurring the costs of color printing.

Some omission error is evident in the illustrative case of Fig. 4.4 since narrow rivers have not been included in the water surfaces category. This is a compromise made in order to avoid commission error that would give cloud shadows the appearance of water surfaces.

4.5 Mixed Mapping Methods

Many multiband mapping methods draw definite distinctions between 'supervised' and 'unsupervised' approaches (Gonzalez and Woods, 1992; Pratt, 1991). The basics of the supervised scenario have been set forth in the foregoing discussion whereby 'training sets' selected on the basis of 'ground truth' serve to determine 'signatures' against which pixel patterns are matched by computational comparison. The unsupervised scenario delays the appeal to ground truth until the later stages of the mapping process. In fact, it is not even necessary to have particular categories in mind at the beginning. Instead, a computational similarity (or dissimilarity) analysis is performed as a clustering operation that segregates the pixels

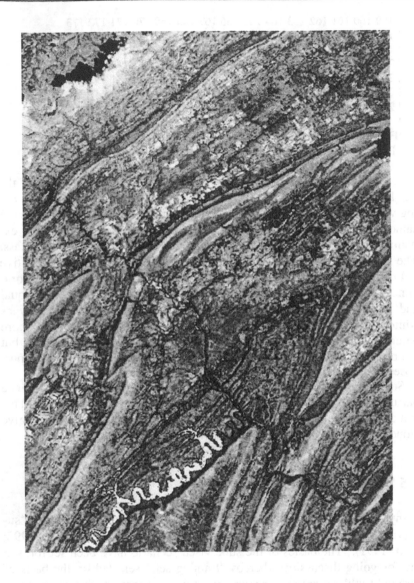

Fig. 4.4 Thematic highlighting of water surface class as white on a gray-tone background for September 1991 Landsat MSS image of central Pennsylvania.

into several groups with the members of any given group having substantial similarity of signal patterns. A sample of each cluster group is then investigated via ancillary information to determine its composition, and appropriate labels are thereby attached. Thus the supervised approach starts with a predetermined legend and some specific knowledge of samples, whereas the unsupervised approach acquires specific information as needed and develops the legend in the course of the investigation.

It can be seen that the pattern-based approaches do not fit neatly under the rubric of either unsupervised or supervised, since ideas from both are incorporated. The pattern-based segmentation with dual levels of detail has the nature of both divisive and agglomerative clustering concepts in statistics. However, the numbers of segments are far more numerous at both levels of detail than is typical of clustering for unsupervised analysis. The coarser A-level of segmentation is akin to what is sometimes called 'hyperclustering' in the context of image analysis, but B-level pattern segments are much more numerous than hyperclustering would entail. The multiplicity of segments invites application of supervised ideas at these levels, with the proviso that distributional assumptions would need to be modified in order to become formally statistical. Thus, the pattern-based approach can be considered in several respects as both hybridizing and extending more conventional approaches.

There is opportunity for pursuing the idea of unsupervised analysis further in the polypattern framework. This would consist of further aggregating the A-level pattern segments by clustering. This can be accomplished by importing fields from .MEN file, .CTR file or .BRS and .BRI files into the analytical spreadsheet of a statistical software package that has clustering capabilities. The resulting cluster membership information can then be cast as a thematic transform in a .THM file as if the cluster numbers were thematic class numbers. A companion thematic tonal transfer table as a .TON file is then created to specify how the respective clusters should be colorized. Rather than entirely arbitrary color tones for clusters, it will usually be better to select values for representative patterns from the .BRS and/or .BRI brightness tables.

The approaches considered in this chapter have been extensively tested in the process of mapping generalized land cover for the entire state of Pennsylvania from polypatterns of Landsat Thematic Mapper (TM) data having 30-meter pixels. Such mapping has proceeded through a second cycle over a period of ten years, and the techniques have evolved in the course of their use. The pattern picture images that supported the mappings have also proven to be of considerable interest to the Pennsylvania public, as evidenced by their utilization on PASDA (Pennsylvania Spatial Data Access).

References

Congalton, R. and K. Green. 1999. Assessing the Accuracy of Remotely Sensed Data: Principles and Practices. Boca Raton, FL: Lewis Publishers/CRC Press. 137 p.

Forman, R. T. T. 1995. Land Mosaics: The Ecology of Landscapes and Regions. Cambridge, U.K.: Cambridge Univ. Press. 632 p.

Forman, R. T. T. and M. Godron. 1986. Landscape Ecology. New York: John Wiley & Sons. 619 p.

Gonzalez, R. C. and R. C. Woods. 1992. Digital Image Processing. Reading, MA: Addison-Wesley Publishing Co. 716 p.

James, M. 1985. Classification Algorithms. London, UK: John Wiley & Sons. 211 p.

McGarigal, K. and B. J. Marks. 1995. FRAGSTATS: Spatial Pattern Analysis Program for Quantifying Landscape Structure. General Technical Report PNW 351, U. S. Forest Service, Pacific Northwest Research Station. 122 p.

Pratt, W. K. 1991. Digital Image Processing. New York: John Wiley & Sons. 698 p.

Tso, B. and P. Mather. 2001. Classification Methods for Remotely Sensed Data. New York: Taylor and Francis. 352 p.

Turner, M., R. Gardner and R. O'Neill. 2001. Landscape Ecology in Theory and Practice: Pattern and Process. New York: Springer-Verlag, Inc. 401 p.

Wilson, J. and J. Gallant, Eds. 2000. Terrain Analysis: Principles and Applications. New York: John Wiley and Sons, Inc. 479 p.

5 Comparative Change and Pattern Perturbation

Focal purposes in research for the polypattern approach to image-structured information have been landscape-level habitat assessment, land cover mapping and landscape change detection. The study of landscape dynamics from an ecosystem health perspective has been the area of greatest emphasis (Myers and Patil, 1999; Patil et al., 2000). Correspondingly, this is one of our 4C concerns for landscape images: *contrast, content, context* and *change*. Land cover and changes in land cover are among the more readily observed broad-scale indicators of status and trends in ecosystem health (Myers et al., 1999). Forests contrast in many respects to grassland/herbaceous environments in terms of habitat characteristics, with savannah and shrub also having special intermediary characters. Forests are constrained by climate in some ecological settings, but moist and mesic environments typically have herbaceous and shrubby areas being disturbance-induced transitional stages in succession to tree cover. Deforestation, reforestation, and fragmentation by land conversion to agriculture and urban have major implications for ecosystem health at landscape and regional scales. Even apart from human influence, however, forest landscapes have a natural disturbance regime due to wind, wildfire, and biotic causes such as insect infestations that create patchy openings of varying sizes in the general matrix of forest cover (Baker, 1989). The most insightful environmental monitoring is aimed at tracking changes to determine whether there are changes in the change regime that would indicate progressive human disruption of ecosystem processes with consequent loss of habitat integrity/connectivity and/or effects of possible global climate change (Patil et al., 2001).

With regard to studying landscape dynamics by means of remote sensing, there are also changes in appearance due to atmospheric conditions and phenology that are not informative relative to alteration of ecosystem status. Clouds and their shadows along with fog and snow constitute major atmospheric sources of interference for monitoring disruption of ecosystem processes. Saturated soil, standing water, and sediment from recent heavy precipitation can substantially alter the spectral reflectance of many environmental surfaces. Phenology pertains to annual cycles of senescence and growth in vegetation. Deciduous forests shed their leaves during dormancy, giving rise to very different appearance between dormant season and growing season. Likewise, fields having agricultural crops change gradually from being bare after tillage to closed green herbaceous plant cover, and then to senescence upon maturity and thence to harvest. To this intrinsic dynamic of agricultural areas is added change due to crop rotation whereby a given field is planted to different crops in different years. Thus, agricultural areas tend to exhibit strong spectral changes even when the nature of human influence on the land is not changing. With taller vegetation and/or topographic relief, there are also substantial localized changes in appearance during the day and between months at a given time of day that are due to shadows and sun angle. Furthermore, some important changes in land cover entail much more pronounced contrast in spectral appearance than others. Finally, temporal comparisons cannot be conducted effectively without making whatever spatial adjustments are needed so that the images and/or maps will overlay each other accurately (Dai and Khorram, 1998; Gong et al., 1992). There are, therefore, copious challenges in characterizing cover change (Coppin and Bauer, 1996; Rogan et al., 2003).

5.1 Method of Multiple Mappings

The approaches presented previously are pertinent to the problem of detecting differences. A seemingly simple solution would be to compare categories of cover maps compiled consecutively. This approach avoids potential phenological problems with shifting spectral signatures, since the classification of a category is conducted with a separate set of spectral signatures or segments in each scene. It also accommodates interim improvements in sensor systems and differing spectral signal segregation in band breakpoints. Consistent categories and mapping methods are essential, however, in order to avoid confounding comparative computations of change.

There has been a propensity of purpose in land cover compilation for funding of focus on current condition and particular perspective. A different perspective produces differences in definitions of categories or even different types of categories that serve to confuse comparisons over time. Although there have been efforts to standardize land cover categories (Anderson et al., 1976), these have not led to the level of consensus needed for long-term monitoring. A further complication with comparative analysis of change in cover mappings is that classification errors from the individual maps combine as apparent change that compounds error in the result (Lunetta and Elvidge, 1998). Therefore, the seemingly straightforward comparison of companion mappings has somewhat limited utility suited to situations where the mappings are done identically and mapping errors are minimal. However, cover mappings have broader utility for landscape change analysis when used in conjunction with spectrally comparative approaches.

5.2 Compositing Companion Images

The general alternatives to comparing cover maps are to compare or composite spectral characteristics of companion images acquired at different times (Mas, 1999; Rogan, Franklin and Roberts, 2003; Singh, 1989). The compositing approach is fundamentally different from the comparison approach. A multi-temporal composite intermixes information on stability of cover types with information on changes, whereas temporal comparison segregates information on change from information on stability of cover types.

A (multi)-temporal composite is obtained by 'stacking' image datasets from different times together so that each pixel has all of the signal bands from the different dates as if the images were all acquired by a single sensor. For the cover types that have not changed between dates, the bands should combine as an extended signature. However, each combination of change types should express in a particular manner that differs from either of the parent types. Since it would be very problematic to locate advance examples for all possible combinations of cover type changes as training sets, an unsupervised approach becomes more practical than a supervised approach. The multiplicity of combined bands also complicates both data management and analysis. This provides incentive for data reduction without sacrificing major elements of pattern information.

Principal component (PC) transformation is a common approach to data reduction that is independent of analytical mode (Byrne et al., 1980; Gong, 1993; Li and Yeh, 1998). The principal component method begins by seeking a linear combination of bands having a form as in Eq. 5.1,

$$Y = c_1 b_1 + c_2 b_2 + \dots + c_n b_n \qquad (5.1)$$

where c is a coefficient and b is a band value for a pixel, such that Y is the projection of the pixel on a rotated dimensional axis having maximum variance among all rotational projections. The method then proceeds to find a second rotational projection that has maximum variance subject to the constraint that it be uncorrelated with the first. The process continues for additional orthogonal (uncorrelated) rotational projections up to the number of original bands. The rotational projections (or dimensions) are called *principal components*, and a subset of the first few often has most of the information content of the original image dataset. A reduced set of principal components is rescaled for encoding in the manner of an image, and used instead of the larger original dataset.

When principal components from a composite image are being used for change detection, considerable caution must be exercised in doing any discarding aimed at data reduction. Changes that are relatively rare in the scene may contribute little to the overall variability in the scene, and consequently tend to appear on the low order principal component axes that would usually be discarded in data reduction. Therefore, each principal component (PC band) should be examined carefully as an individual image. One should discard only those PC bands that appear as random speckle lacking in pattern. Principal components concepts can also be applied to pattern properties as set forth in the next chapter.

Polypattern segmentation can serve a purpose similar to principal components in analysis of temporal composites. The stacked bands are subjected to segmentation in the usual manner of a multiband image. There will tend to be isolation of changes in some of the A-level patterns, but not necessarily as especially light ones or especially dark ones. Therefore, the initial gray-tone image from direct rendering of the polypattern overtones will tend not to be very revealing relative to changes, and it becomes necessary to work with various color renderings in order to make the changes evident. Fig. 5.1 shows a pattern picture of the composited 1972 and 1991 Landsat MSS sub-scenes of central Pennsylvania. Careful perusal will show tonal distinctions that are drawn from both of the individual scenes, but these distinctions have subtle gray-tones instead of highlights. An exception to the subtlety is the appearance of cloud effects from both scenes in the upper left corner. Color is required in order to bring out the other change aspects that have been captured. An effective colorizing strategy in this case is to show combined visible signals as red, pattern brightness order as green, and combined infrared (somewhat dampened) as blue.

5.3 Direct Difference Detection

Although colorized multi-date composites can provide additional landscape details, they have the disadvantage of not definitely distinguishing

Fig. 5.1 Gray-scale rendering of A-level polypatterns for multi-temporal composite of Landsat MSS sub-scenes from 1972 and 1991 in central Pennsylvania.

change from stability. This ambiguity makes it necessary to investigate each color signature empirically to determine whether it is indicative of change. Lack of definite records on changes in the localities under investigation can make determination of change signatures difficult. Direct spectrally comparative indicators of change are thus of interest for use either with multi-date composites or separately (Howarth and Wickware, 1981).

If images having the same set of spectral bands are available on successive dates, then an obvious possibility is to look at differences between the respective pairs of bands (Bruzzone and Prieto, 2000). Particular kinds of differences will be most evident for certain bands, so that it may not be sufficient to examine the difference between just one band pairing. Therefore, it becomes desirable to incorporate differences for multiple band pairs into a single indicator of change. Euclidean distance between signal vectors on different dates is a natural extension of differencing from one band to several bands. This follows from the fact that difference and Euclidean distance are numerically identical when there is only one band. In remote sensing terminology, the Euclidean distance between matched signal vectors on two dates has usually been called 'length of change vector' as the basic indicator in *change vector analysis* (Chen et al., 2003; Johnson and Kasischke, 2003; Lambin and Strahler, 1994a & 1994b; Michalek et al., 1993).

Considering their very much localized nature and the possibility of mixed-pixel effects, there is generally not much interest at the landscape level in single-pixel changes. With regard to changes of a patchy nature, subtle variations among pixels in a patch constitute noise in what should be interpreted as a patch of change. Therefore, homogenizing of patches can serve to give a more crisp appearance to change indicator images. This is exactly the smoothing effect that is introduced by polypattern segmentation. Thus, the polypattern version of change vector length is to calculate the Euclidean distance between the proxy property points for A-level patterns from different dates of imagery. A software module for this purpose is introduced in Appendix B.8. Atmospheric anomalies such as clouds and their shadows are sources of false change (Hall et al., 1991; Song et al., 2001). The module makes provision for designating lists of A-level patterns that are not to be considered as being change. The methods of the preceding chapter can be used to determine which patterns should be ignored in this way.

The computed change vector lengths are scaled into 255 classes, with the classes being numbered 1 through 255 and 0 indicating missing data. The mean vector length and pixel count for each class are reported in a special version of a .BRS brightness file. The class intervals are controlled by a scale factor. The scale factor is a multiplier for the mean change vector length. This multiple of the overall mean length is scaled equally into the first 254 classes, and anything larger appears in class 255. Making the scale factor smaller will make smaller spectral changes more evident, whereas making the scale factor larger will tend to suppress smaller spectral changes. There is also an option to set a threshold for class number such that vector lengths in smaller classes will not be indicated as change in a pattern pictures. Choice of scaling and thresholds is usually done empirically, but can become a subtopic of investigation in change detection (Fung and LeDrew, 1988).

The basic output of the direct difference detector is an image lattice containing class numbers for change vector lengths with optional suppression of classes less than a threshold. There is also an option to produce a multiband image lattice including the individual band components of change vectors as additional bands. The change vector length then becomes the last band. If any of these band components has a difference with magnitude larger than 125, then all components are scaled so that the magnitude of the largest is 125. Scales for components are then shifted by the amount of 125 to avoid having negative values. Thus, a component value of 125 represents an actual difference of zero, and components less than 125 are actually negative.

The polypattern lattices to be compared for change must have the same directional orientation and the same pixel resolution (size), but they need not have identical areas of coverage. Coupling of the two images is accomplished by specifying row and column positions in each that mark the same location. The coverage of the output lattice is the same as that of the first one, with all non-overlapping positions being given a zero value. Fig. 5.2 shows pattern-based direct difference detection by change vectors for the Pennsylvania Landsat MSS sample scenes with darker tones indicating stronger change. The changes are much more apparent when rendered in color with red for change and blue for stable areas.

Depending upon the purpose, it may not always be appropriate to include all bands in the change vector computations, especially in the case of thermal bands. Thermal bands can give the appearance of change due to heating and cooling of environmental materials even in the absence of actual change in land cover.

It can be seen from Fig. 5.2 that a major feature of change is the large sinuous Raystown Lake reservoir in the lower portion of the area. Construction for this impoundment was underway at the time of the 1972 image, and the area was under water in 1991. The concentrations of agricultural fields in the valleys also appear as areas of change in signal pattern, even though the nature of the land use is unchanged. This underscores the

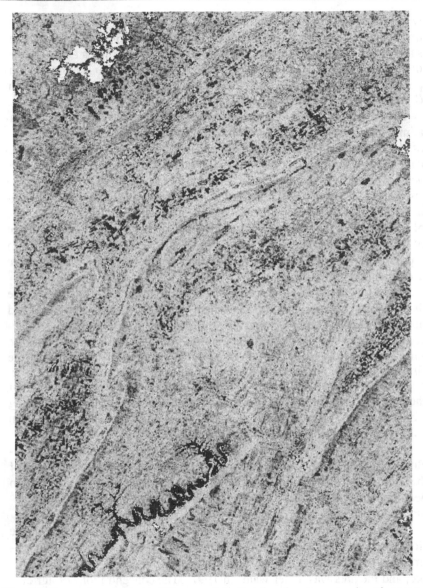

Fig. 5.2 Pattern-based direct difference detection by change-vector length as an indicator of landscape change from 1972 to 1991 using Landsat MSS data for central Pennsylvania, with darker tones indicating change. The more dense clouds and cloud shadows have been excluded from change, and thus both appear as white.

distinction between land use and land cover. The land cover has changed, but the land use has not. Since the land cover change is real, this is not a difference that is subject to removal by clever signal processing. Therefore, an understanding of the landscape dynamics for the areas is important to intelligent interpretation.

5.4 Pattern Perturbation

The conventional change vector and the foregoing pattern-based equivalent are both constrained to usage when the same signals have been obtained on all occasions of data acquisition. Neither multiple mappings nor temporal composites are constrained in this manner, which appears to give them a broader scope of application. The polypattern process of segmentation, however, entails a duality of signal and spatial information that can be exploited to lift the constraint in an innovative manner. The essence of the innovation lies in analysis of the manner in which patterns are perturbed over time. The key to the method is inter-date matching for pattern counterparts. This counterpart approach focuses on consistency and inconsistency of spatial organization for patterns between dates, thus avoiding the presumption that signal acquisition has been conducted in the same manner.

One A-level of polypatterns for a pair of occasions is chosen as the *base occasion* for comparison. Let the kth A-level pattern for the base occasion be denoted as $\downarrow PP_k$ and let $\uparrow PP_j$ denote the jth A-level pattern for the other occasion to be linked (*link occasion*) by pattern matching. Pattern $\downarrow PP_k$ occupies a particular set of pixel positions for the base occasion. These same positions are scanned for the link occasion to determine the pattern $\uparrow PP_j$ that occurs most frequently in this subset of pixels. This modal pattern becomes the counterpart of $\downarrow PP_k$ which we denote as $\downarrow PP_k \blacktriangleright \uparrow PP_j$. Every pixel for the link occasion thus has two patterns associated with it. One of these is its own pattern $\uparrow PP_i$ and the other is the pattern $\uparrow PP_j$ expected to occur there on the basis of pattern matching $\downarrow PP_k \blacktriangleright \uparrow PP_j$ from the base occasion. Euclidean distance in signal space between $\uparrow PP_i$ and $\uparrow PP_j$ is then computed as a change vector length. Note that both of these patterns are from the signal information of the link occasion. Since it is the spatial organization of patterns that drives the matching, signal information from the base occasion plays only an indirect role. If the spatial organization of patterns is the same for both dates, then the two signal patterns will be the same. If the spatial organization of patterns is consistent but not identical, then the two signal patterns will differ relatively little. Consistency may mean that some areas of coarser and finer subdivision have been exchanged between dates. If a pixel was associated with different regions having contrasting signal characteristics, however, then the two patterns will also differ considerably. In effect, a pixel is judged as being consistent or not by the company that it keeps in its patterns on the two dates.

Fig. 5.3 is a change vector mapping from this indirect counterpart strategy for comparison with Fig. 5.2 that was produced by the direct strategy. For this purpose, the earlier image was designated as the base and the later one as the linkage. With this methodology it is possible to do either forward comparisons in time or backward comparisons in time. A backward comparison would involve the later image as base and the earlier as linkage. Forward and backward comparisons will not necessarily yield exactly the same results.

As described in Appendix B.9, the pattern matching adds an extra step to the change vector determination scenario as compared to the direct computation of segment-based change vectors described previously.

There is an indeterminacy to such spatially based comparisons that imparts robustness to some changes of imaging conditions and sensors. If a forest clusters consistently in leaf-on and leaf-off conditions, for example, the phenological difference will not register as a landscape change. If a landscape clusters consistently with different sensors, there will likewise not be confounding with temporal changes. There is, however, a subtle possibility for masking certain localized landscape changes (Ramakomud, 1998; Beck, 1999). If there are contrasting 'islands' limited to a particular surrounding cover type in the base image, then absence of such islands on the link image will not be detected.

This indirect method for comparing sets of image-structured environmental indicators has utility outside the immediate domain of remote sensing. The habitat diversity mappings introduced in the first chapter provide an example. Habitat diversity was mapped in terms of both species richness and regional habitat importance index (RHII). It is of interest to inquire how these two perspectives on habitat diversity may lend emphasis to different areas. Fig. 5.4 shows the result of using the indirect method of segment counterparts to make such a comparison. Differences in local emphasis come from occurrence of numerous common species versus fewer species that are of particular conservation concern. Lighter portions of Fig. 5.4 are areas of agreement where both views show either low diversity or high diversity. Darker areas are areas of disagreement where one view indicates low or high diversity in conflict with the other view.

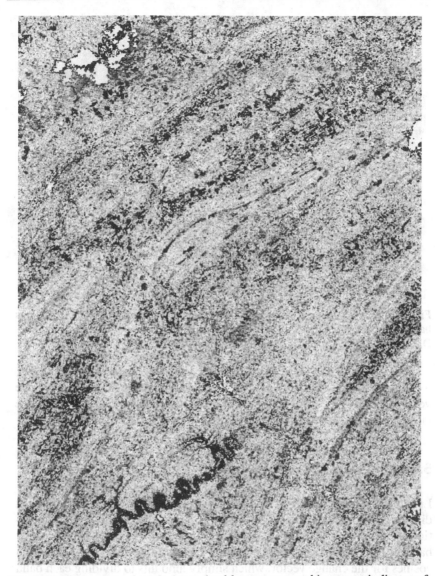

Fig. 5.3 Change-vector length determined by pattern matching as an indicator of landscape change from 1972 to 1991 using Landsat MSS data for central Pennsylvania, with darker tones indicating change.

Fig. 5.4 Indirect comparison by pattern counterparts of habitat diversity mappings based on species richness as opposed to regional habitat importance index (RHII). Darker areas have greater difference in local emphasis.

The capability is also available for mapping components of the change vectors in the manner of bands, and for imposing thresholds on change vector mappings. Such components can be used to look beyond presence or absence of change to investigate differences in the nature of changes.

5.5 Integrating Indicators

Length of change vector is readily interpreted as an image, but does not discriminate between differences in spectral direction of change. Information on directions of change is embodied in a multi-layer dataset of scaled band differences. One possible scaling in this regard is that of direction cosines for the change vector, which simply amounts to dividing each band component by the length of the change vector (Chen et al., 2003). From an image display perspective, however, a more convenient scaling is one that centers and stretches the band differences to fit a byte range of 0 to 255 as shown in Fig. 5.5 for the band 3 difference component. The real question is how to handle these multiple layers of information in combination so that distinctions in types of changes can be made and seen as images.

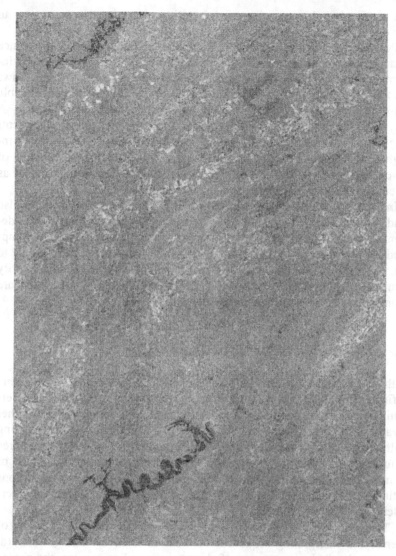

Fig. 5.5 Image of scaled band 3 component of change as computed by pattern matching from Landsat MSS of central Pennsylvania from 1972 to 1991. Notice that the change for flooding of reservoir is seen differently from changes in vegetative cover land surface.

The pattern-based approach to segmentation also provides an answer to this question.

Polypattern segmentation can be conducted on multiband difference data in like manner as for original multiband data. This will combine information on direction of changes into a single image-map that allows pseudo-color depiction of directional differences. There is considerable scope for creative coloration in rendering such complex composites. A gray-tone view of A-level patterns combining band differences and vector length is given in Fig. 5.6. However, a colorized view is much more informative. A good strategy for colorizing in this case is to have length of the change vector appear as red, the total of changes for infrared bands as green, and the total of visible band changes as blue.

In working with polypatterns of multiple change indicators it is also prudent to obtain the residual map as described in Chapter 2 in order to determine whether there are locations of strong change that have been dampened in the approximation. Fig. 5.7 shows the residuals corresponding to Fig. 5.6. In this case the larger (darker) residuals are associated primarily with clouds and with the Raystown Lake reservoir, both of which are strongly expressed in the segmentation itself.

5.6 Spanning Three or More Dates

With conventional approaches to change detection there is considerable difficulty in tracking across three or more occasions for purposes of determining persistence of changes in extended landscape monitoring. Another advantage of the polypattern approach is that the concept of segmenting multiple indicators extends quite naturally to analysis of change for three or more dates. The simplest version of this is working with a stack of change vector images for successive pairs of occasions. Color renditions of the A-level change patterns will show different aspects of temporal persistence in different colors.

Change vectors between successive pairs of occasions are naturally of interest and interpreting sequences of these is relatively straightforward. For more detailed investigation of shorter temporal sequences, the choice of change vectors to be included in the stack can become more sophisticated. Consider image data for three occasions (O_1, O_2 and O_3) with polypattern representations in terms of change vector length. In addition to comparisons between successive pairs of occasions (O_1-O_2 and O_2-O_3), also conduct comparison O_1-O_3 between beginning and ending dates. The latter comparison will have some redundancy with those for successive

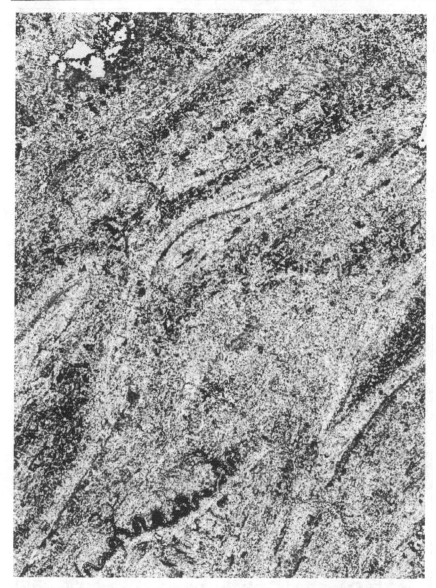

Fig. 5.6 Gray-tone view of A-level patterns combining band differences and vector length from Landsat MSS of central Pennsylvania between 1972 and 1991.

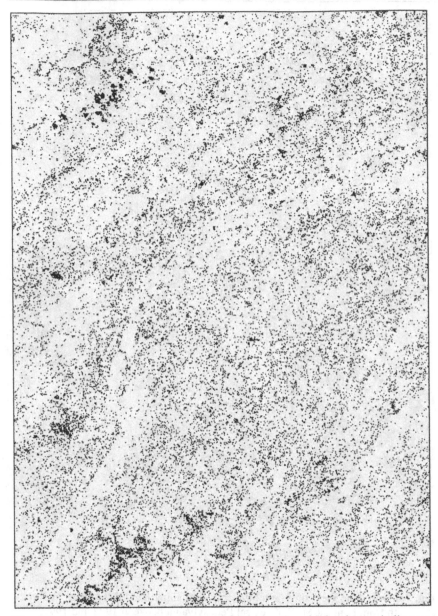

Fig. 5.7 Image of residual map for polypatterns of multiple change indicators shown in Fig. 5.6, with darker areas indicating larger residuals.

dates. The more persistent changes occurring in either O_1-O_2 or O_2-O_3 will also appear in O_1-O_3. Shorter duration O_1-O_2 changes, however, will not necessarily appear in O_1-O_3. The scaling options for change vectors should be consistent for all paired comparisons, and a threshold should be used that is the same throughout. With lengths of change vectors stacked in the order O_1-O_2, O_2-O_3, O_1-O_3, it is suggested that the color rendition for the composite use O_1-O_3 as red, O_2-O_3 as green, and O_1-O_2 as blue. This uses red for changes from beginning to end, green for later changes, and blue for earlier changes. The composite can also provide an approximated view for any particular date pair by placing the same pairing of occasions on all three colors to generate a gray-tone image.

The combination of change vector lengths for successive pairs of occasions with those for selected longer spans can be extended to sequences of more than three occasions. Comparisons from beginning to end and middle to end are normally in order. Need for temporal resolution and timing of events that are of special interest can guide additional choices.

References

Anderson, J., E. Hardy, J. Roach and R. Witmer. 1976. A Land Use and Land Cover Classification System for Use with Remote Sensor Data. U. S. Geological Survey Professional Paper 964. Reston, VA: U. S. Geological Survey. 28 p.

Beck, F. 1999. Cluster Counterparts and Echelons in Remote Sensing Change Detection. Master of Science Thesis in Forest Resources, The Pennsylvania State University, Univ. Park, PA 16802. 138 p.

Bruzzone, L. and D. Prieto. 2000. Automatic Analysis of the Difference Image for Unsupervised Change Detection. *IEEE Transactions on Geoscience and Remote Sensing* 38(3): 1171-1182.

Baker, W. 1989. A Review of Models of Landscape Change. *Landscape Ecology* 2: 111-133.

Byrne, G., P. Crapper and K. Mayo. 1980. Monitoring Land-Cover by Principal Component Analysis of Multitemporal Landsat Data. *Remote Sensing of Environment* 10: 175-184.

Chen, J., P. Gong, C. He, R. Pu and P. Shi. 2003. Land-Use/Land-Cover Change Detection Using Improved Change Vector Analysis. *Photogrammetric Engineering and Remote Sensing* 69(4): 369-379.

Coppin, P. and M. Bauer. 1996. Digital Change Detection in Forest Ecosystems with Remote Sensing Imagery. *Remote Sensing Reviews* 13:207-234.

Dai, X. and S. Khorram. 1998. The Effects of Image Misregistration on Accuracy of Remotely Sensed Change Detection. *IEEE Transactions on Geoscience and Remote Sensing* 36(5): 1566-1577.

Fung, T. and E. LeDrew. 1988. The Determination of Optimal Threshold Levels for Change Detection Using Various Accuracy Indices. *Photogrammetric Engineering and Remote Sensing* 54: 1449-1454.

Gong, P. 1993. Change Detection Using Principal Component Analysis and Fuzzy Set Theory. *Canadian Journal of Remote Sensing* 19(1): 22-29.

Gong, P., E. LeDrew and J. Miller. 1992. Registration Noise Reduction in Difference Images of Change Detection. *International Journal of Remote Sensing* 13(4): 773-779.

Hall, F., D. Strebel, J. Nickeson and S. Goetz. 1991. Radiometric rectification: Toward a Common Radiometric Response Among Multidate, Multisensor Images. *Remote Sensing of Environment* 35(1):11-27.

Howarth, P. and G. Wickware. 1981. Procedures for Change Detection using Landsat Digital data. *International Journal of Remote Sensing* 2(3): 277-291.

Johnson, R. and E. Kasischke. 1998. Change Vector Analysis: a Technique for the Multispectral Monitoring of Land Cover and Condition. *International Journal of Remote Sensing* 19(3): 411-426.

Lambin, E. and A. Strahler. 1994a. Indicators of Land-Cover Change for Change Vector Analysis in Multitemporal Space at Coarse Spatial Scales. *International Journal of Remote Sensing* 15: 2099-2119.

Lambin, E. F. and A. H. Strahler. 1994b. Change-Vector Analysis in Multitemporal Space: a Tool to Detect and Categorize Land-Cover Change Processes Using High Temporal-Resolution Satellite Data. *Remote Sensing of Environment* 48: 231-244.

Li, X. and A. Yeh. 1998. Principal Component Analysis of Stacked Multi-Temporal Images for Monitoring of Rapid Urban Expansion in the Pearl River Delta. *International Journal of Remote Sensing* 19: 1501-1518.

Lunetta, R. and C. Elvidge, eds. 1998. Remote Sensing Change Detection: Environmental Monitoring Methods and Applications. Ann Arbor, MI: Ann Arbor Press. 350 p.

Mas, J. 1999. Monitoring Land-Cover Changes: a Comparison of Change Detection Techniques. *International Journal of Remote Sensing* 20: 139-152.

Michalek, J., T. Wagner, J. Luczkovich and R. Stoffle. 1993. Multispectral Change Vector Analysis for Monitoring Coastal Marine Environments. *Photogrammetric Engineering and Remote Sensing* 59: 381-384.

Myers, W., G. P. Patil and C. Taillie. 1999. Conceptualizing Pattern Analysis of Spectral Change Relative to Ecosystem Status. *Ecosystem Health* 5(4):285-293.

Patil, G. P., R. Brooks, W. Myers, D. Rapport and C. Taillie. 2001. Ecosystem Health and Its Measurement at Landscape Scale: Toward the Next Generation of Quantitative Assessments. *Ecosystem Health* 7(4): 307-316.

Patil, G. P. and W. Myers. 1999. Environmental and Ecological Health Assessment of Landscapes and Watersheds with Remote Sensing Data. *Ecosystem Health* 5(4):221-224.

Patil, G. P., W. Myers, Z. Luo, G. Johnson and C. Taillie. 2000. Multiscale Assessment of Landscapes and Watersheds with Synoptic Multivariate Spatial

Data in Environmental and Ecological Statistics. *Mathematical and Computer Modelling* 32: 257-272.

Ramakomud, A. 1998. Change Detection Using Hyperclustered Data: the Spatial Averaging Approach. Master of Science Thesis in Electrical Engineering. The Pennsylvania State University, Univ. Park, PA 16802. 189 p.

Rogan, J., J. Miller, D. Stow, J. Franklin, L. Levien and C. Fischer. 2003. Land-Cover Change Monitoring with Classification Trees Using Landsat TM and Ancillary Data. *Photogrammetric Engineering and Remote Sensing* 69(7): 793-804.

Rogan, J., J. Franklin and D. Roberts. 2002. A Comparison of Methods for Monitoring Multitemporal Vegetation Change Using Thematic Mapper Imagery. *Remote Sensing of Environment* 80(1): 143-156.

Singh, A. 1989. Digital Change Detection Techniques Using Remotely Sensed Data. *International Journal of Remote Sensing* 10(6): 989-1003.

Song, C., C. Woodcock, K. Seto, M. Lenney and S. Macomber. 2001. Classification and Change Detection Using Landsat TM Data: When and How to Correct Atmospheric Effects. *Remote Sensing of Environment* 75: 230-244.

6 Conjunctive Context

Among the four *4C* focal concerns of image analysis (*contrast, content, context* and *change*) the one yet to be considered in detail is *context*. Several preliminary comments about this aspect have been introduced in previous chapters, but its complexity and the disparity among avenues of approach have made it appropriate for further consideration to await treatment of the other three. This topic encompasses the spatial structure and positioning of the patterns in the lattice and speaks to spatial statistics. We resume reviewing residuals from the close of Chapter 2 where concern was regarding restoration, but now considering the contextual complement with residuals as results of detrending.

6.1 Direct Detrending

Spatial statistics is a special sub-discipline that concerns itself with the manner in which spatially referenced data vary when differences in location are explicitly taken into account (Schabenberger & Gotway, 2005). Although we have the information necessary to locate all instances of a pattern in the spatial lattice, we have not as yet actually utilized that location information in any of the analysis except for showing image and pattern information in pictorial form. Aside from the possibility of some file ordering effects on our pattern processes, we could have obtained essentially equivalent results from a randscape (random landscape). In particular, it is clear that we have not yet addressed any co-location aspects (in a proximity sense) of the image information; whereas these aspects are a major focus of spatial statistics. We have essentially separated the non-spatial from the spatial aspects of image information. This separation is intentional, because it allows us to contribute to the field of multivariate spatial analysis in a way that does not contaminate any of its outcomes.

Statistical analysis usually entails background assumptions of various sorts, but many of the approaches in spatial statistics impose assumptions that are especially rigorous. Typical of these is an assumption of at least weak stationarity for the signals in the lattice. This assumption implies that the mean of the signal is constant over the extent of the lattice, and that any tendency of two things closer together to be more alike does not depend on where the things are situated in the lattice but only on the separation of their positions. There is often a further assumption of isotropy, which implies that there are no pronounced directional effects. It should be apparent from the images we have already showed that these assumptions are highly unrealistic for most actual landscapes.

The strategy for resolution of these incompatibilities is one of *detrending*, whereby all of the apparent spatial trends in the lattice are modeled and then removed by subtraction to leave a trendless lattice of residuals. Spatially statistical analysis is then conducted on the detrended lattice of residuals. With complex signal surfaces like those of landscapes, however, it is not a statistically straightforward matter to model the trends in order to subtract them. This is especially so for modeling of joint trends in multivariate (multiband) lattices. Here is where polypatterns have a role to play.

The A-level patterns and the B-level patterns are both spatially specific, and thus provide models of the joint trends among the signals in the lattice but with different degrees of fidelity. Since both pattern levels are also approximate relative to the original image data, they will leave a lattice of residuals upon subtraction. The A-level serves to show the general nature of the trends, whereas the B-level should be effective for filtering out the trends by subtraction. If residual images show evidence of remaining trends, then additional splitting cycles can be conducted for further refinement.

Fig. 6.1 revisits the residual information of Fig. 2.4 which was presented without procedural particulars of processing. The residuals are determined as Euclidean distance between the proxy signal vector for the B-level pattern and the actual signal vector for the particular pixel. Appendix B.7 introduces a facility for determining and mapping multivariate residuals of this nature. Both the original image and the polypatterns are needed as inputs for determining the residuals.

The computed residual distances for the pixels are scaled into 255 classes, with the classes being numbered 1 through 255 and 0 indicating missing data. The mean residual distance and pixel count for each class are reported in a special version of a .BRS brightness file. Note carefully that this version of the .BRS file is not intended for use in composing images of the residuals. This special version of the .BRS table has three columns. The first column is residual class number, the second is average residual distance for the class, and the third is number of pixels belonging to the class. The residual class intervals in the .BRS table are controlled by a scale factor. The scale factor is a multiplier for the mean residual distance

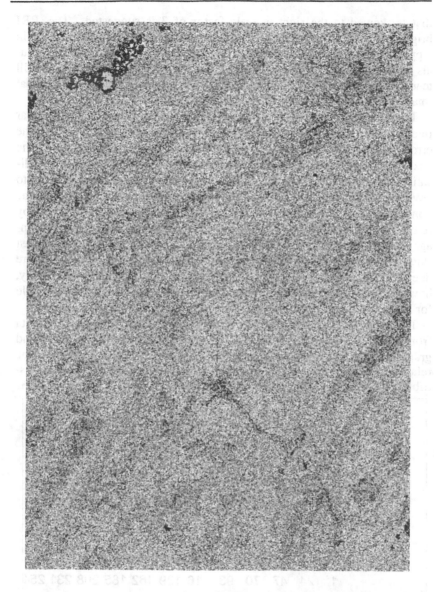

Fig. 6.1 Multi-band residual image for September 1991 Landsat MSS image of central Pennsylvania with darker tones indicating larger residuals.

of the image as a whole. This multiple of the overall mean residual distance is scaled equally into the first 254 classes, and anything larger appears in class 255. Since approximation errors tend to have a distribution that is skewed to the right, a scale factor of 3 is suggested. The .BRI brightness table is used for making images of the residuals as described in Appendix B.7. Making the scale factor smaller will make smaller residuals more evident in the image, whereas making the scale factor larger will tend to suppress smaller residuals. In this sense, appearance of the residual image is relative to choice of scale factor.

The spatial structure or lack thereof in the residual image is of particular interest. An 'ideal' pattern would be one of uniformity, indicating that the errors of approximation were evenly distributed over the image area. The next most favorable pattern would be a random 'speckle' distribution indicating that the errors of approximation constitute 'white' noise relative to environmental features and locations. Typically, however, there is some nonrandom spatial patterning of the residuals that can be investigated by reference to pattern pictures. Then it becomes necessary to refer to the image itself to determine what kinds of environmental features have the least fidelity in their representation. The stronger discrepancies in this case are due to the spectral and spatial complexity of cloud fringes. Certainly, however, stationarity assumptions would be considerably more reasonable for Fig. 6.1 than for the original in Fig. 1.1.

To study the distribution of residuals by statistical graphics, the data from the .BRS file can be imported into a statistical spreadsheet and graphed in the manner of a histogram. Although the class values will be relative, the shape of the distribution will nevertheless be evident. A distributional graphic of this nature is shown in Fig. 6.2.

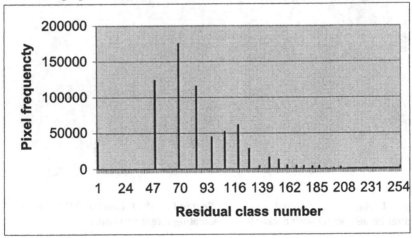

Fig. 6.2 Distributional chart for residual classes mapped in Fig. 6.1.

6.2 Echelons of Explicit Spatial Structure

Spatial structure of a surface can be expressed both explicitly and empirically in terms of *echelons*. Echelons extract spatial structure as hierarchies of hills or topological territories (Myers, 2005; Myers & Patil, 2002; Smits & Myers, 2000; Myers, Patil & Taillie, 1999; Myers, Patil & Joly, 1997).

Echelons are recursive arrangements of peaks, foundations of peaks, and foundations of foundations forming a topological tree. Saddles are the surface structures that separate echelons from each other as shown schematically in Fig. 6.3.

Simple peaks constitute one order of echelons, foundations for peaks formed at saddles constitute another order of echelons, and foundations for foundations formed at subsequent saddles constitute higher orders of echelons. The immediately overlying echelon structures supported by a foundation comprise its progeny, with the foundation being parent at a node in a topological tree.

The order relations among echelons are like those used in characterizing drainage networks of streams. A peak is first order. A foundation for two or more simple peaks is second order. A foundation for two second orders is third order, and so on with order increasing at a junction of like orders. A suite of attributes is compiled for each echelon, including basal extent and relief as change in height from top to base along with order and parent-progeny relations.

Our usual intuitive model for echelons is one of receding floodwaters with peaks emerging as islands followed by progressive merging of islands to form larger islands. Compound islands represent families of echelons. It is at the structural level of a family, or branch of the topological tree, that an echelon complex must necessarily constitute a connected component of the surface.

There is an indeterminacy of echelons by which concavities in a surface that would "hold water" are not identified. This indeterminacy can be resolved by the echelon structure of the complement surface for which hills correspond to valleys in the original surface.

Fig. 6.4 shows the first-order echelons (peaks down to but not including first saddle) for the residual surface of Fig. 6.1, after leveling single-pixel spikes and holes. This has the appearance of a noise surface, except for the location of the cloud bank in the upper-left corner. In this 1000×700 grid, there are 43,303 first-order echelons after the single-pixel filtering. This extreme abundance of small peaks is also indicative of a noise surface.

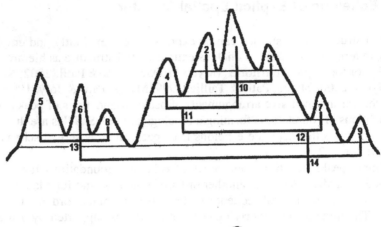

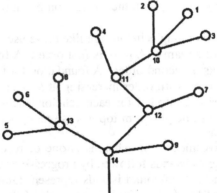

Fig. 6.3 Echelon formation by surface saddles and consequent hierarchical relations.

It is computationally cumbersome to obtain higher-order echelons of a noisy surface such as this. In this case, first-order echelons serve our purpose of showing the lack of sustained trend over most of the mapping extent.

Spatial structure in models of images can likewise be explored through echelons, but each component of a signal vector must be analyzed separately as a scalar surface. Similarities of spatial structure can then be studied for the respective bands as signal surfaces.

Fig. 6.4 Order 1 echelons (light) of residual surface from Fig. 6.1 after filtering single-pixel spikes and holes.

6.3 Disposition and Situation

To continue with context, a pattern PP_k in a lattice L is *distributed* over a set of *positions*, which we will call the *disposition* of the pattern in the lattice. Context for a pattern is inherently conjunctive since it entails relation to the part of the lattice (if any) that is not in the disposition of the pattern, which we symbolize by $\neg PP_k$ as the negation of the pattern in L. Relative to the polypatterns of the 1991 Landsat MSS sub-scene of central Pennsylvania, Fig. 6.5 shows the disposition of PP_{101} which is the most prominent of the patterns by virtue of occupying about 4.2% of the lattice. It is readily apparent that this particular pattern is highly dispersed despite its prominence.

We take advantage of our classification work in Chapter 4 to compare the disposition of PP_{101} with the disposition of PP_6 in Fig. 6.6 which we know from content to imply water. The disposition of PP_6 is obviously much more localized than that of PP_{101}, although not necessarily as connected components. We introduce a qualitative descriptor to express such obviously different characters of disposition. We say that PP_{101} is *weakly situated* whereas PP_6 is *strongly situated*. The logic behind this terminology is that the nature of landscapes is such that a pattern like PP_6 will occur in a fairly definite landscape setting or situation that is quite readily subject to further specification upon more detailed examination. To the contrary, it may require a more extensive investigation to characterize the circumstances associated with occurrence of PP_{101} and the resulting characterization may be less certain.

6.4 Joint Disposition

Consider also the disposition of PP_7 as shown in Fig. 6.7. Both the disposition and the situation of PP_7 appear very similar to PP_6. Visual comparison suggests that we try picturing the *joint disposition* of PP_6 and PP_7 as the union of the sets of positions occupied by these two patterns as shown in Fig. 6.8. It is clear that the joint disposition is even more strongly situated than for either of the component patterns. A question that naturally arises from this observation is whether further generalization of patterns would be appropriate to facilitate making inferences regarding landscape organization. This is a question which we temporarily defer.

We also take note of three interesting *pattern propensities* with regard to joint disposition. As observed with PP_7 and PP_6, the patterns may be jointly *integrative* such that the combination is more strongly situated than

Fig. 6.5 Disposition of PP$_{101}$ (light) with ¬PP$_{101}$ (dark) for September 1991 Landsat MSS sub-scene of central Pennsylvania.

Fig. 6.6 Disposition of PP_6 (light) with $\neg PP_6$ (dark) for September 1991 Landsat MSS sub-scene of central Pennsylvania.

either of the members. Another interesting propensity is when the patterns are *transitional*, whereby the joint disposition extends the area occupied by either member individually. Third is the *gradational* case where the patterns both interleave and extend.

The usual approach for analysis of context by landscape ecologists is through *patch metrics* (McGarigal and Marks, 1995) that presume a mosaic structure consisting largely of instances of *connected components* imbedded in a *matrix*. Such metrics are typically based on patch areas, patch perimeters and patch shapes or some combination of these, including elongation or stringiness as expressed, for example, in fractal dimension.

These examples should be sufficient to illustrate that polypatterns may not fit this underlying concept of patch structure, instead being in the nature of what we might call *proto-patchiness* having substantial structural components of singles, doubles, triples, etc. Under these circumstances the usual patch metrics are more or less degenerate and ineffective at capturing features of dispositions and joint dispositions of patterns. This does not imply lack of utility for patch metrics, but rather that the application pertains to certain types of *meta-patterns* as patterns of polypatterns to be considered subsequently.

If one is willing to temporarily set aside the quantitative characteristics of signal vectors, however, the proto-patchiness and/or partial patchiness of dispositions shown above can be addressed by methods of spatial statistics for *marked point processes on a lattice* as described by Schabenberger & Gotway (2005). We refer the reader to that source and references cited therein, rather than attempting to summarize their exposition.

6.5 Edge Affinities

One aspect of joint disposition which is not hindered by proto-patchiness is that of edge affinities whereby there is preferential sharing or avoidance of edges with another pattern relative to a randscape having the same number of pixels in each pattern. For example, from visual examination of joint distribution in Figs 6.5 – 6.8, it should not be surprising that PP_6 and PP_7 share more edges in common than expected if their positions were randomized. What may be surprising is that PP_7 is not the strongest partner for PP_6 in this regard, but rather PP_3. Fig. 6.9 exhibits the disposition of PP_3 and Fig. 6.10 shows the joint disposition of $\{PP_6 \cup PP_3\}$. Thus, it should be evident that analysis of edge affinities among patterns is useful not only for context, but also for content by suggesting what additional patterns might accompany a training set for a particular landscape

Fig. 6.7 Disposition of PP_7 (light) with $\neg PP_7$ (dark) for September 1991 Landsat MSS sub-scene of central Pennsylvania.

Fig. 6.8 Joint disposition of {PP$_6$ ∪ PP$_7$} (light) with ⫴{PP$_6$ ∪ PP$_7$} for September 1991 Landsat MSS sub-scene of central Pennsylvania.

Fig. 6.9 Disposition of PP$_3$ (light) with ¬PP$_3$ (dark) for September 1991 Landsat MSS sub-scene of central Pennsylvania.

Fig. 6.10 Joint disposition of $\{PP_3 \cup PP_6\}$ (light) with $\neg\{PP_3 \cup PP_6\}$ for September 1991 Landsat MSS sub-scene of central Pennsylvania.

category that is expected to exhibit strong patchiness as is the case for water surfaces.

PSIMAPP software support for the polypatterns includes a facility for analysis of edge affinities among A-level patterns. This facility is introduced in Appendix B.10. An edging profile is computed for each A-level pattern. The first part of the profile compares the order of prominence for the pattern with its edge order when self-edges and outside edges are excluded. The next part of the profile gives the fraction of self-edges for the pattern. Then the numbers of edge patterns to attain deciles of edges are listed. After the deciles, there is a listing of edge patterns that exhibit greatest affinity relative to the criterion of randomness. A companion file gives complete edge affinity ratings for all patterns relative to all other patterns.

6.6 Patch Patterns and Generations of Generalization

We return now to the previously deferred topic of grouping patterns into meta-patterns (patterns of patterns) as Figs. 6.6 – 6.10 would suggest. The underlying issue is that patterns as we have defined them and derived them are informational constructs rather than entities of landscape organization. The A-level polypatterns are configured to accommodate the constraints of a byte of computer storage, and the B-level patterns reflect computational cutoffs. The evidence in Figs. 6.3 – 6.10 indicates that segmentation even for A-level patterns has gone beyond conditions of connectedness in that landscape. Experience with other landscapes suggests that this is commonly true.

An obvious first thought would be to import the .CTR table of central values into a statistical spreadsheet having capability for clustering, and then proceed in the usual manner with agglomerative grouping. However, this leaves open the question of when to cease aggregating. It would be highly desirable to have an aggregation process for which cessation is self-determined and that also respects the spatial structure of the patterns.

A possible strategy is to merge patterns that are nearest neighbors in signal space if a neighbor is among the k highest with respect to edge affinity. This strategy is self-limiting given prior choice of k, and takes account of signal/spatial dualities of patterns. The joining of patterns having high degree of edge affinity will necessarily increase the patchiness, thus being appropriately called *patch patterns*.

If it still appears that further generalization is appropriate, there are at least two avenues available. One is to relax the k constraint somewhat.

The other is a next 'generation' of aggregation after first providing an appropriate proxy signal vector for each composite. This new generation will then be similarly self-limiting at a new level of generalization. This latter is quite different from the usual agglomerative approaches whereby greater aggregation is obtained as a single sequence of increasing dissimilarity that is judgmentally truncated, since it increases spatial connectedness in the patterns and limits judgments to number of generations.

Such generalization must not interfere with the integrity of the polypattern encoding. This can be accomplished by replacing segment pattern vectors by vectors for patch pattern proxies in copies of the respective tables for the polypatterns.

Running a first generation patch-pattern process on the September 1991 Landsat MSS scene with mergers limited to the 15 (6%) highest edge affinities has the effect of consolidating the 250 A-level patterns to 231, for which the main consequence is joining of water-related patterns. This is shown as an image in Fig. 6.11.

6.7 Parquet Polypattern Profiles

One relatively straightforward way of generalizing contextual analysis like that of edge affinities is to tabulate frequencies of co-occurrence for pattern pairs in blocks of pixels as vicinities. Because of boundary issues, a sliding window could reveal more than abutting windows. Even though a sliding window entails counting a pattern in a particular pixel multiple times, the relative counts are still suitable for indexing to random expectation in parallel manner as for edge information. Unfortunately, sliding windows produce much larger datasets.

An innovatively different approach is to partition the lattice of patterns into square block parquet subsets such as 10×10 pixels or 15×15 pixels and tabulate the distribution of pattern frequencies in each block as a *pattern profile* for the block. For block size of 15×15 or smaller, the pattern counts could be recorded in one byte of computer storage for each A-level pattern in a profile. This gives a coarser lattice of image-structured data in which the respective pattern counts play the role of signal bands. It then becomes possible to conduct higher-order contextual analysis from the pattern profiles. Tricolor images can be developed to show salient characteristics of the profiles. One such approach is to use the segment number marking 30% of the cumulative frequencies as blue, the segment marking 60% of the cumulative frequencies as green, and the 90% one as red.

Fig. 6.11 Pattern picture showing first generation generalization of patch-patterns for September 1991 Landsat MSS image of central Pennsylvania.

Profiles consisting entirely of low numbered segments would then appear as dark grayish with tinge of orange, and those consisting entirely of high numbered segments as light gray to whitish with a tinge of orange. Different mixes of low and high numbered segments would appear in particular colors. While avoiding the cost of color illustrations here, Fig. 6.12 shows a gray-tone view for segment number at 90% cumulative frequency in 10x10 blocks of the September 1991 Landsat MSS scene for central Pennsylvania. The general landscape patterns of high and low signal intensities are evident at this scale of compilation.

The profiles lend themselves to a kind of supervised classification of landscapes according to pattern structure. Profiles for landscape types of interest can be extracted to serve as training sets, and then computationally matched against other areas according to profile (dis)similarity. For purposes of assessing (dis)similarity, we also propose a new measure of distance between two pattern profiles that we call *accordion distance*.

Let $\square_{Ai}$ be the frequency of occurrence of the ith pattern in profile A and $\square_{Bi}$ be the corresponding frequency of occurrence of the ith pattern in profile B. Also, let $\mathcal{C}_{Ai}$ be the cumulative frequency for pattern numbers $\leq i$ when pattern number i is ≤ 125; and the cumulative frequency for pattern numbers $\geq i$ when pattern number i is ≥ 126. Then squared accordion distance $\mathcal{D}^2$ is given by Eq. 6.1.

$$\mathcal{D}^2{}_{AB} = \Sigma(\mathcal{C}_{Ai} - \mathcal{C}_{Bi})^2 \qquad (6.1)$$

Two-byte encoding for profile counts relaxes block size constraints by accommodating blocks of 250×250 pixels. It then becomes possible to conduct multi-scale contextual analyses by collapsing blocks into blocks of blocks so that thresholds of extent for different landscape settings or situations can be determined empirically.

6.8 Conformant/Comparative Contexts and Segment Signal Sequences

The multiple bands or signals underlying image patterns may have some or substantial redundancy in the sense of covariation and correlation. Although covariation and correlation are familiar statistical features, it is less well appreciated that the correspondence among signals may shift between the units that give rise to the signals. In the case of the electromagnetic spectrum, this entails differences in the spectral specificity of environmental materials. Some environmental materials are quite uniformly re-

Fig. 6.12 Gray-tone representation of segment number at 90% cumulative frequency in 10x10 blocks of September 1991 Landsat MSS scene for central Pennsylvania.

flective/absorptive across substantial ranges of wavelengths in the spectrum, in which case we refer to them mainly in terms of their lightness or darkness. Accordingly, newly fallen snow appears light to the eye since it is highly reflective across the visible range of wavelengths. On the other hand, colored environmental materials differ substantially among wavelengths in their absorption and reflection characteristics.

At least with respect to spectral signals, therefore, it should not be difficult to envision that degree of correspondence is specific to landscape context. Thus, covariance and correlation among such signals across extensive scenes can be different according to the mixes of environmental materials comprising the scenes. Therefore, it is informative in a landscape sense to segregate those segments that exhibit conformance among the signals so that study of comparative characteristics of the signals can be focused on the remainder. The concept of partial ordering (Patil & Taillie, 2004) can be extended to serve the purpose of determining where and in what degree the signals are conformant.

The first step is to establish the degree to which the respective segments have strong or faint signals. The matter of contrast stretching or otherwise enhancing signals has been considered in a previous chapter. Such enhancement makes the scaling of the signals somewhat arbitrary, so it is appropriate to begin by expressing the strength of each signal in each segment as a rank. For present purposes, we use the 'place' way of ranking in which the segment having the highest value for a given signal (band) has rank one (first place), the second highest value is rank two, etc. Comparative analysis of strong or faint is then conducted in terms of these ranks. Thus, a high rating implies faintness of the signals.

A segment can be considered as having signal strength if it has at least one signal band that is ranked at least as strong as for any other segment. Such a segment is not 'dominated' by any other segment. All such segments receive a signal strength rating level of one, where again this 'first place' is seen as having superior strength of signal. The remaining segments are then considered apart from the segments having the level one designation. One or more among these remaining segments will not be dominated by any of the other remaining segments, and these are assigned a signal strength rating level of two. The process then continues recursively for unrated segments until all segments have received a strength rating level. As the rating level increases, there is increasing consensus on overall faintness of signals for a segment. Thus, the ratings are best considered by increasing numerical order as a composite 'faintness' indicator.

Among the stronger-signal segments (low faintness value), there are also those for which one or more of the signal bands are fainter. Therefore, the next concern is to consider the range of ranks for signal strengths within the segments. Polypatterns of the September 6, 1972 Landsat MSS shown in Figs. 3.2 – 3.4 are advantageous for illustrating this kind of contextual analysis due to senescence of the herbaceous vegetation with consequent conformant reflectivity across the bands. Fig. 6.13 shows rank range plotted against faintness level for the segments of this image. The faintest segments have a small range of ranks as expected, but there is little clarity to be seen for the segments having stronger signals. Therefore, further analysis to extract usable structure in this regard becomes necessary.

As a step toward extraction of structure, it is suggestive to graph rank statistics for segments as vertical lines. The bottom of the vertical line for a segment marks the minimum signal rank, the top marks the maximum signal rank, and the median rank is marked in the line. This type of graph is shown in Fig. 6.14 for the segments of the September 6, 1972 Landsat MSS scene of central Pennsylvania. Examination of the graph reveals an overall trend in the ranks from low numbered segments to high numbered segments, but the departures from the trend are quite erratic. Extraction of structure will consist of re-sequencing the segments to obtain several series having consistent progression of ranks within each series. A progression will be considered consistent if each successive segment has a minimum that is greater than or equal to the minimum of the preceding segment and also a maximum that is greater than or equal to the maximum of the preceding segment. There is a regular algorithm for accomplishing this re-sequencing.

The first operation in re-sequencing is to reorder the segments according to increasing minimum rank, with sub-ordering of any ties according to maximum rank, then sub-ordering of ties according to median rank. A consistent series is then obtained by dropping out any segment that is followed by one having a lesser maximum. A consistent series is likewise extracted from the (ordered) dropouts, with additional series being extracted recursively until the dropouts are themselves a consistent series. These series are called *rank range runs* (Myers et al., 2006), and a number is assigned to each run in the order of extraction. An overall numbering is also assigned to segments that spans the entire set of rank runs. Fig. 6.15 is a vertical line graph of rank runs by overall segment sequence number. With reference to a segmented scene, it is appropriate to describe the rank range runs as *segment signal sequences*. It can be readily observed that the segment signal sequences toward the left side of the graph have narrower ranges, with extremely wide ranges occurring at the right side of the graph. Therefore, the leftward series represent conformant contexts wherein the different signal bands are responding more or less alike.

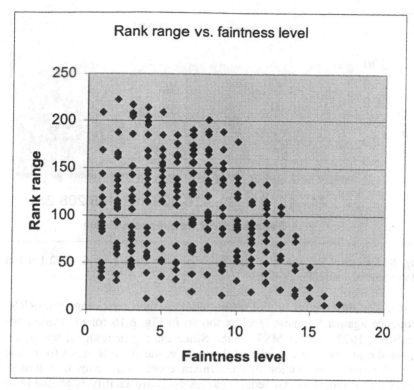

Fig. 6.13 Rank range of signals on Y-axis versus faintness level on X-axis for segments of 4-band Sept. 6, 1972 Landsat MSS image of central Pennsylvania.

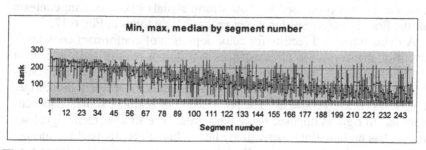

Fig. 6.14 Vertical line from min rank to max rank with median mark for each segment of September 6, 1972 Landsat MSS scene of central Pennsylvania.

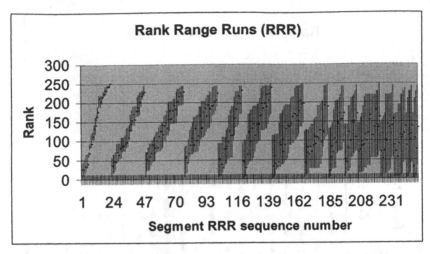

Fig. 6.15 Vertical line graph of rank range runs for September 6, 1972 Landsat MSS scene of central Pennsylvania.

It is also informative to plot the segments in their rank range run (RRR) sequence against faintness level as shown in Fig. 6.16 for the same September 6, 1972 Landsat MSS scene. Since the segment signal sequences are also quite evident in a plot of this nature, the stage is set to formulate rules for special coloration of conformant contexts in terms of faintness levels and segment signal sequences. A software facility is available to modify a color transfer table for pattern pictures according to specifications (rules) given in terms of faintness level and segment signal sequence. That facility has been used to show strong signals in conformant contexts for the first five segment signal sequences as light tones in Fig. 6.17.

A more automated facility for color depiction of conformant contexts is also available for use in circumstances that do not require specific control of the coloration. This is a tricolor approach that uses green to red gradation for strong to weak signals in terms of faintness of level, with intermediate levels being yellowish. The progression of segment signal sequences from left to right is imposed using blue tints. The first segment signal sequence has no blue tint. Increasingly heavy blue tints are used for successive segment signal sequences. With increasing blue tint, red will grade to magenta, green will grade to cyan, and yellow will grade to gray. Accordingly, these latter gradations indicate progressively less conformance among the signal bands.

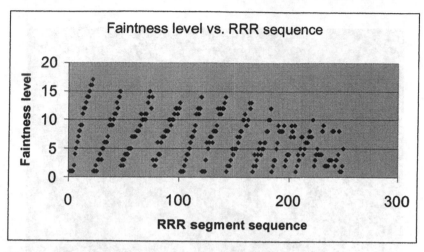

Fig. 6.16 Plot of faintness level against RRR segment sequence showing segment signal sequences for September 6, 1972 Landsat MSS scene of central Pennsylvania.

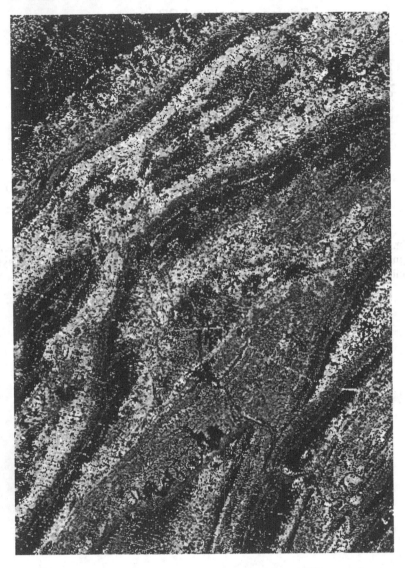

Fig. 6.17 Pattern picture of September 6, 1972 Landsat MSS scene in central Pennsylvania showing senescent herbaceous vegetation in the valleys as light toned conformant contexts.

6.9 Principal Properties of Patterns

Contextual relations among the signal bands in relation to the patterns can also be investigated from a different perspective through an adaptation of principal components. Principal components were considered briefly in Chapter 5 with respect to change detection. As introduced then, principal component analysis resolves covariances or correlations into non-redundant linear combinations of the original variables (signals), while retaining the full range of variation present in the data from which covariances or correlations were calculated.

The basic input to a principal component analysis consists of either a covariance matrix or a correlation matrix. A software facility for computing overall covariances and correlations for the original image data is introduced in Appendix B.11, and another facility gives covariances and correlations from the A-level information. Extracting principal components from a covariance matrix can show how the respective signal bands contribute to each of the principal components, and thereby also the extent to which signals have redundancy by virtue of contributing heavily to the same principal components. Rendering the principal components as images will enable visual examination of the factors having informational independence among multiple signal bands. The first couple of principal components represent factors that contribute most to the variability in the image. Subsequent components contribute progressively and often considerably less to the informational composition. It may happen, however, that certain components having low overall contribution can highlight locally restricted but interesting features.

One question with regard to polypatterns that does not arise in conventional image analysis concerns the degree to which the correlation structure among pixel primitives is captured in the A-level segments. This is also one of the quality considerations with respect to efficacy of the segmentation. Table 6.1 shows the correlation matrix for the September 1991 Landsat MSS scene in central Pennsylvania at the level of pixel primitive data. Table 6.2 shows the corresponding correlation matrix as computed directly for the A-level segments alone. It is apparent from comparison of these two tables that the correlation structure of the A-level segments is very similar to that of the pixel primitive data. In particular, the two signal bands in visible range of spectral wavelengths are highly correlated and the same is true for two signal bands in the infrared range of wavelengths. However, the correlations are much less between the visible and the infrared signal bands. This supports the statements made in the early chapters

Table 6.1 Band-to-band correlation matrix for pixel primitive data of the September 1991 Landsat MSS scene in central Pennsylvania.

	Band 1	Band 2	Band 3	Band 4
Band 1	1.000000	0.968536	0.559148	0.373645
Band 2	0.968536	1.000000	0.518148	0.316423
Band 3	0.559148	0.518148	1.000000	0.955278
Band 4	0.373645	0.316423	0.955278	1.000000

Table 6.2 Band-to-band correlation matrix for A-level segment data of the September 1991 Landsat MSS scene in central Pennsylvania.

	Band 1	Band 2	Band 3	Band 4
Band 1	1.000000	0.965185	0.501907	0.320052
Band 2	0.965185	1.000000	0.453183	0.255167
Band 3	0.501907	0.453183	1.000000	0.957354
Band 4	0.320052	0.255167	0.957354	1.000000

that information from infrared wavelengths is complementary to information from visible wavelengths.

Since the correlation structure at the A-level is consistent with the pixel primitive level, it is reasonable to conduct principal component investigations at the A-level using statistical spreadsheets instead of resorting to specialized image analysis software packages. Accordingly, the band-to-band covariance matrix for the A-level is shown in Table 6.3 as a basis for computing principal components.

Table 6.3 Band-to-band covariance matrix for A-level segment data of the September 1991 Landsat MSS scene in central Pennsylvania.

	Band 1	Band 2	Band 3	Band 4
Band 1	28.809557	34.359886	25.519003	19.788759
Band 2	34.359886	43.989212	28.472076	19.495224
Band 3	25.519003	28.472076	89.731506	104.466072
Band 4	19.788759	19.495224	104.466072	132.696884

The first principal component accounts for 78.24% of the variation in the image data and the second component accounts for 20.67%, leaving only about 1% for the two remaining components together. Fig. 6.18 shows an image of the second principal component. Examination of this image indicates that the second principal component is primarily in the nature of a contrast between heavily vegetated areas and areas where vegetation is sparse or absent.

Fig. 6.18 Image of second A-level principal component for September 1991 Landsat MSS image of central Pennsylvania.

The effectiveness of principal components for A-level segments bodes favorably for similar experimentation with other kinds of transformational multivariate analyses at this level. Thus, it should be evident that contextual analysis is a very expansive and fertile field for research into polypatterns obtained from image-structured information.

References

Mirkin, B. 2005. Clustering for Data Mining. Boca Raton, FL: Chapman & Hall/CRC, Taylor & Francis. 266 p.

Myers, W. 2005. Surface Structures and Tiered Topography. *Environmetrics* 16:1-13.

Myers, W and G.P. Patil. 2002. Echelon Analysis. Pp. 583-586 in *Encyclopedia of Environmetrics*, Volume 2. A. El-Shaarawi and W. W. Piegorsch, eds. UK: Wiley.

Myers, W, G.P. Patil and K. Joly. 1997. Echelon Approach to Areas of Concern in Synoptic Regional Monitoring. *Environmental and Ecological Statistics*, 4:131-152.

Myers, W., G.P. Patil and Y. Cai. 2006. Exploring Patterns of Habitat Diversity Across Landscapes Using Partial Ordering. In: R. Bruggemann and L. Carlsen, eds. Partial Order in Environmental Sciences and Chemistry. Springer. 406 p.

Myers, W, G.P. Patil and C. Taillie. 1999. Conceptualizing Pattern Analysis of Spectral Change Relative to Ecosystem Status. *Ecosystem Health*, 5:285-293.

Patil, G.P. and C. Taillie. 2004. Multiple Indicators, Partially Ordered Sets, and Linear Extensions: Multi-criterion Ranking and Prioritization. *Environmental and Ecological Statistics* 11, 199-228.

Schbenberger, O. and C. Gotway. 2005. Statistical Methods for Spatial Data Analysis. New York: Chapman & Hall/CRC. 488 p.

Smits, P and W. Myers. 2000. Echelon approach to characterize and understand spatial structures of change in multitemporal remote sensing imagery. *IEEE Transactions on Geoscience and Remote Sensing*, 38(5):2299-2309.

7 Advanced Aspects and Anticipated Applications

This concluding chapter has two primary purposes. One purpose is summary and synthesis of salient strengths for polypatterns produced by segmentation sequences. The other is to preview some of the advanced aspects that are being addressed in recent research or are anticipated for future focus. We first seek synthesis of the foregoing features to set the stage for additional analytical opportunities.

7.1 Advantageous Alternative Approaches

Our segmentation scenarios provide alternative approaches with analytical advantages as compared to conventional image investigation. The analytical advantages arise primarily from inducing explicit entities that are otherwise inherently implicit in the image information. Particularly for nearly natural landscapes, image information involves geographic gradients that often elude explicit extraction. Our segmentation scenarios systematically select subsets of signals as explicit entities implying heuristic hypotheses regarding presence of pattern in pictures. These explicit entities have emergent properties that are open to many modes of purposeful pursuit. Since the entities are extracted entirely in the signal domain, they are subject to investigative validation in the spatial domain.

There are seven notable areas of advantage in our approach. One of these lies in controlling contrast of pictorial presentations. Whereas conventional computations of contrast enhancement are per-pixel propositions, we compute A-level transfer tables of picture properties for segments that produce palettes for patterns. Depictions are done indirectly by developing columns in the transfer tables using pattern properties instead of the slower scanning of scenes. This also supports segment selection for special suppression or showcasing. Investigation of innovative integration of signals as image indicators can be done in spreadsheet software instead of complex computational configurations of pixel processing packages.

A second aspect of advantage in our approach pertains to classifying content for constructing categorical maps. Whereas digital image data are usually directed toward algorithmic assignment of image elements to candidate categories of content, our approach is equally applicable to assisting interpretive assignment by a human analyst. With a suitable interactive interface, the analyst can pick a particular pattern and examine its disposition over the image area in order to use landscape logic for deductively deciding how to identify it in relation to the structure of the scene. Segment signal similarities and proximate positioning can also be incorporated into the interpretive investigation instead of being treated as a mutually exclusive mode of mapping. The translation tables for pictorial presentation can also be integrated into the mapping scenario so that particular patterns can be shown as categorical color imbedded directly into tricolor renderings of the surrounding scenery.

A third aspect of appreciable advantage lies in detecting differences between instances of imaging. The classic case is for sequences of scenes taken over time with the same sensor. In this relatively simple scenario, the signal smoothing from pattern proxies helps to make perturbed patches prominent. Additional advantage and innovative improvement comes from matching mosaics of patterns in paired instances of imaging. Perturbation of patterns in matched mosaics provides difference detection even with a shift in sensing systems. Segregating sets of signals from a sensor also allows assessment of differences in detectors. Still more substantial is the second-order advantage available by polypattern processing after compositing indicators of change across multiple instances of imaging. The patterns of patterns in change reveal temporal trajectories in landscape dynamics.

The most pervasive aspect of advantage lies in contextual considerations, which also enter into the content and change aspects addressed above. Having parsed patterns into collective components allows analysts to conduct comparatives in multiple modes. The components can be combined according to signal similarities and proximate positioning to generate generalized images that portray progressively more prominent patterning. The patterns can be treated as multivariate trends for removal to reach residuals that are regionalized in accordance with scenarios of spatial statistics. The separate signals can be seen as surfaces having structures in terms of echelons that reveal relationships in terms of topological territories. An entire new arena of analysis is posed by pattern profiles of cumulated components over blocks of pixels at several spatial scales for locating landscapes. Compositional components of complexes can be considered in terms of chromaticity or ratio relations among signal sets by partial ordering and rank range runs.

A fifth aspect of advantage is informational compression for conveyance by computer media. The polypattern models occupy the equivalent of two single-byte bands along with ancillary tables of pattern properties. Restoration is approximate, which has the effect of a smoothing filter. Maps of the distribution of residuals can be made at the time of pattern preparation, and refinements made to the models if appropriate. Therefore, at least three signal bands are needed to create compression. The more bands there are, the greater the compression.

Although approximation in restoration might appear to be a drawback, it leads to the sixth aspect of advantage. Digital image data from satellite sensors are often purchased under contractual or copyright restrictions on redistribution. Since the polypattern models do not provide capability for complete restoration, and in view of their numerous advantage aspects as currently considered, they become substantially different derivative products in much that same manner as a thematic map. The A-level model can, in fact, be considered as thematic map of landscape patterns. Therefore, most of the proprietary concerns relative to the original data should be largely obviated.

The seventh aspect of advantage lies at the interface between image analysis and Geographic Information Systems (GIS). GIS provides the popular platform for utilization of geo-spatial information. Relatively few of the regular GIS users are also image analysts. As pointed out immediately above, the A-level image model of polypatterns can be treated in the manner of a thematic layer taking the form of a raster map. This confers compatibility with the more common GIS. The information would become image-like again through B-level restoration, but that will not be necessary for many purposes in a GIS user environment. Polypattern packaging thus facilitates broader access to image-based information.

7.2 Structural Sectors of Signal Step Surfaces

Echelon expression of surface structure was introduced in chapter 6 for the purpose of showing that the model accounted for most of the topological trends in the data, since the profusion of minute peaks in the residuals distributed throughout the area did not accommodate any substantial sustained trends. Of equal or greater interest in most instances is the structure of topological trends that have been captured in the model. Such structures are usually examined in the A-level model of polypatterns, since the B-level model typically encompasses more detail than can be effectively absorbed in a visual mode or even with supporting statistical displays.

This aspect of analysis is of interest particularly for working with image-structured environmental indicators, since these can be more intuitively comprehended as surfaces than as images. When tabulated over a grid of pixels, such indicators form step surfaces rather than smooth surfaces. The stepping is particularly pronounced with integrative image indicators such as a normalized difference vegetation index (NDVI), whereas some sense of smoothness is imparted by interpolation of data originally available only from particular points. In our suite of examples, the biobands of diversity in habitat suitability for vertebrates are illustrative of environmental indicators.

It is interesting that the most obvious sense of surface comes from each indicator comprising a single signal 'band' of information. Accordingly, the tables of pattern properties allow each indicator to be restored with some smoothing that is akin to interpolation at the A-level for elucidation of surface structure as orders, extents and families of echelons comprising topological territories. However, the making of the model by pattern processes also brings forward the novel multivariate modeling that transpires in polypatterns. The modeling is done jointly among indicators rather than individually, and the ordered overtone indexing of the A-level model is a composite expression of the indicators. Thus, treating the A-level overtone indexes directly as a surface of signal synthesis can be done as one way of investigating multivariate topological trends among the indicators.

The 'bioband' dataset of diversity in vertebrate habitat suitability serves to show the six-surface synthesis of signals as depicted earlier in Fig. 2.2, which can be compared visually to individual indicators such as mammals in Fig. 1.4 and fish in Fig. 1.5. It is instructive to extract and explore echelons of the ordered overtones to exemplify this as an advanced analytical approach in multivariate methodology.

An initial issue is to consider whether single-pixel peaks in such a surface are of any interest, since they are often numerous and of a noisy nature. Therefore, we invoke an echelon option for shaving off single-pixel spikes and filling single-pixel pits so that larger-scale surface structure becomes more evident. This constitutes a special sort of spatial filter function. There are 2945 echelons even after filtering spikes and pits, so the composite surface is relatively complex with 6 being the highest order of the echelons. An echelon order image of the surface is shown in Figure 7.1, with first-order echelons being brightest and brightness decreasing as echelon order increases. The aquatic groups exert a strong influence since the drainages are prominent.

We have focused our echelon exploration here on a multivariate model because prior applications of echelons have been largely limited to simpler surface scenarios. Echelon analysis has been applied to individual

Fig. 7.1 Image of echelon orders for overtones of vertebrate species richness in terms of habitat suitability for Pennsylvania, with order 1 being brightest and order 6 being darkest.

biodiversity layers (Myers et al., 1997; Johnson et al., 1998; Myers and Patil, 2002; Myers, 2005). Echelons of spatial structure in change indicators from imagery have been investigated quite extensively (Myers et al., 1999; Beck, 1999; Smits & Myers, 2000; Liu, 2000). Among other image indicators, the normalized difference vegetation index (NDVI) as also been studied in terms of echelons (Kurihara et al., 2000; Ricotta & Avena, 2000). Topological concepts underlying echelons have been extended as *dome domains* in ecological mapping (Myers et al., 2005). Echelon inspired tree-structured representations of surfaces have been developed for purposes of hotspot detection in geoinformatics (Kurihara & Kong, 2002; Kurihara, 2003; Ishioka & Kurihara, 2005; Patil et al., 2004; Patil & Taillie, 2004; Myers et al., 2006).

7.3 Thematic Tracking

There is considerable contemporary concern with keeping land cover maps current, and also with characterizing changes in land cover. These two concerns are conventionally treated as being different, and Chapter 5 contains considerations for capturing change while Chapter 4 addresses adaptive approaches for classifying cover in thematic mapping. Imaginative integration of the coverage in Chapters 4 and 5, however, presents possibilities for coupling the two concerns.

These possibilities hinge on the pattern matching of mosaics that is mentioned in Chapter 5 for picking out perturbed patches. If mosaic matching is used to point on probable perturbation, it should also serve simultaneously for seeking stable sites. Given a cover classification from the initial instance of imaging, the stable sites should serve as training sets for specifying signatures in the second scene. One could classify the second scene using this suite of signatures to obtain an updated map. Then the categories of the current classification could be compared to those of the earlier map in order to elucidate exchanges of categories over the observational interval. This would require that the current categories be carried over from the previous period. The sequence of steps for such thematic tracking would thus be the following:

1. Maintain the mapping for the initial instance of imaging.
2. Perform polypattern processing on the initial imagery and on the second scene.
3. Use mosaic matching for finding places of probable perturbation for capturing change.
4. Modify mosaic matching to seek stable sights.
5. Systematically select signatures from stable sights.
6. Classify categories of current cover.
7. Match previous map and modified map in patches of probable perturbation to elucidate exchanges of categories, thus showing specifics of change.
8. Investigate inconsistencies and correct categories or changes.

Such a sequence of steps could be adaptively automated for maintaining maps, monitoring landscape dynamics, and revealing change in change regimes relative to hypothesized natural disturbance regimes.

7.4 Compositional Components

Conformant contexts and segment signal series as introduced in Chapter 6 are based on partial ordering analysis and extensions such as rank range runs. These approaches have been applied primarily as pathways to prioritizing entities that show inconsistencies of ordering on multiple indicators (Myers & Patil, 2006). For present purposes, the signal bands are the indicators and the polypatterns are the entities to be ordered. A perspective to be pursued for our purposes is that the entities may not constitute a comparable pool, in which case we would prefer to partition the pool into portions that are relatively regular in their compositional components as reflected in signal ratios. An extension of partial ordering with rank range runs can be conceived as an anticipated application of polypatterns.

Our prospective process views a multivariate measurement vector on an entity as determining a direction in signal space. Compositionally comparable entities will share approximate alignment along a direction in signal space, but with varying lengths of vectors. An initial extraction of rank range runs via Hasse levels of partial ordering could be conducted for considering components that are approximately balanced in signal strength across the bands, thus being effectively achromatic. The K most consistent rank range runs (RRR) would be recorded and removed temporarily from the pool.

For the remaining entities, principal components would pick the entity with the largest projection on the major axis as being prominent in the censored signal space. A transformation vector can then be calculated that will equalize the values of all variates for this entity, and then applied to the entire original pool. The transformation effectively rotates and rescales signal space so that the selected entity becomes achromatic, which is to say that it is situated along the multi-dimensional 45-degree line of shifted signal space.

Consistent rank range runs are again recorded and the associated entities removed in addition to removing those from the previous pass. The process would then repeat until only a few entities remain that can be considered as outliers.

It is anticipated that a strategy of this nature would separate, for example, vegetated components from other types of components in polypattern models of images. It should be noted that principal components is not an essential operation, because it only serves to pick the next type of interesting entity for investigation in a systematic manner. It would also be possible to pursue a similar strategy in which the next type of entity is picked purposively by interactive interpretation.

A question that is left unresolved in our sketch of strategies is how to decide on the number of rank range runs (K) to be considered consistent and thereby removed in each cycle. There are a number of possibilities in this regard, including that of simply exercising retrospective judgment by examining graphical representations of the rank range runs. A further variant would be to reverse the conditioning of Hasse levels so that filtering is done on the basis of signal weakness instead of signal strength.

7.5 Scale and Scope

We have left what are undoubtedly the most elusive aspects of patterns in landscapes to last. These are the aspects of scaling and scope, and they are ones to which we have devoted appreciable attention in our related research. Scope concerns the area extent encompassed by an image, with an attendant 'law of averages' that increasing scope of coverage tends to encompass greater diversity of landscape elements. Scaling properties pertain to rate of degradation in detail with reduction in resolution, which arises from various kinds of averaging or aggregating over increasingly large pixels.

Our previous pursuit of these properties has centered on categorical coverage, particularly with regard to classes of land cover (Johnson & Patil, 2006). We were able to address this aspect effectively in terms of multi-resolution fragmentation profiles and conditional entropy (Johnson et al., 1999; Johnson et al., 2001). We were also able to employ stochastic generating models for simulating hierarchically structured multi-cover landscapes (Johnson et al., 1999). This experience lays a good groundwork upon which to build using the conceptual constructs of polypatterns and parquet profiles of A-level polypatterns.

The polypatterns are quasi-categorical inasmuch as they constitute classes as collectives of pixel properties, with the pixel properties being both quantitative and qualitative. The parquet profiles are frequency functions, but each frequency is associated with a vector of values. Therefore, the polypatterns with their properties and multi-resolution parquet profiles present opportunities for analysis on all levels of quantification. We have thus broken ground in an extremely fertile field of spatial scope and scaling. It opens new frontiers of opportunity for hierarchical modeling in landscape ecology and image analysis. We invite participation in extending the exploration of these dual domains.

References

Beck, F. 1999. Cluster Counterparts and Echelons in Remote Sensing Change Detection. Penn State Univ. Master of Science Thesis. The Pennsylvania State University, Univ. Park, PA.

Ishioka, F. and K. Kurihara. 2005. Detection of Hotspots for Principal Component Space Using Voronoi Regions and Echelon Analysis. Proceedings of the 5th IASC Asian Conference on Statistical Computing, pp. 77-80.

Johnson, G., W. Myers, G. P. Patil and D. Walrath. 1998. Multiscale Analysis of the Spatial Distribution of Breeding Bird Species Richness Using the Echelon

Approach. Technical Report No. 96-1101. Center for Statistical Ecology and Environmental Statistics. Pennsylvania State University, University Park, PA. 19 p.

Johnson, G., W. Myers and G.P. Patil. 1999. Stochastic Generating Models for Simulating Hierarchically Structured Multi-cover Landscapes. *Landscape Ecology* 14, 413-421.

Johnson, G., W. Myers, G.P. Patil and C. Taillie. 1999. Multiresolution Fragmentation Profiles for Assessing Hierarchically Structured Landscape Patterns. *Ecological Modelling* 116, 293-301.

Johnson, G., W. Myers, G.P. Patil and C. Taillie. 2001. Characterizing Watershed-delineated Landscapes in Pennsylvania Using Conditional Entropy Profiles. *Landscape Ecology* 16, 597-610.

Johnson, G. and G.P. Patil. 2006. Landscape Pattern Analysis for Assessing Ecosystem Condition. Boston, MA: Kluwer Academic Publishers. In press.

Kurihara, K. 2003. Detection of Hotspots Based on Hierarchical Spatial Structure. *Bulletin of the Computational Statistics of Japan* 15 (2), 171-183.

Kurihara, K., W. Myers and G.P. Patil. 2000. Echelon Analysis of the Relationship between Population and Land Cover Patterns Based on Remote Sensing Data. *Community Ecology* 1, 103-122.

Liu, S.T. 2000. Response of Echelons to Spatial Patterns of Spectral Change in Pennsylvania Landscapes. Penn State Univ. Master of Science Thesis. The Pennsylvania State University, Univ. Park, PA.

Myers, W. 2005. Surface Structures and Tiered Topography. *Environmetrics* 16, 1-13.

Myers, W., N. Kong and G.P. Patil. 2005. Topological Approaches to Terrain in Ecological Landscape Mapping. *Community Ecology* 2, 191-210.

Myers, W., K. Kurihara, G.P. Patil and R. Vraney. 2006. Finding Upper-level Sets in Cellular Surface Data Using Echelons and SaTScan. *Environmental and Ecological Statistics* 13(4). In press.

Myers, W., G.P. Patil and Y. Cai. 2006. Exploring Patterns of Habitat Diversity Across Landscapes Using Partial Ordering. In: R. Bruggemann and L. Carlsen, eds. Partial Order in Environmental Sciences and Chemistry. Springer. 406 p.

Myers, W., G.P. Patil and C. Taillie. 1999. Conceptualizing Pattern Analysis of Spectral Change Relative to Ecosystem Status. *Ecosystem Health* 5 (4), 285-293.

Myers, W. and G.P. Patil. 2002. Echelon Analysis. Pp. 583-586 in *Encyclopedia of Environmetrics*, Volume 2. A. El-Shaarawi and W.W. Piegorsch, eds. UK: Wiley.

Myers, W., G.P. Patil and K. Joly. 1997. Echelon Approach to Areas of Concern in Synoptic Regional Monitoring. *Environmental and Ecological Statistics* 4, 131-152.

Patil, G.P., J. Bishop, W. Myers, C. Taillie, R. Vraney and D. Wardrop. 2004. Detection and Delineation of Critical Areas Using Echelons and Spatial Scan Statistics with Synoptic Cellular Data. *Environmental and Ecological Statistics* 11 (2), 139-164.

Patil, G.P. and C. Taillie. 2004. Upper Level Set Scan Statistic for Detecting Arbitrarily Shaped Hotspots. *Environmental and Ecological Statistics* 11(2), 183-197.

Ricotta, C. and G. Avena. 2000. Analysis of the Spatial Distribution of Net Primary Productivity Across Corsica (France) Using the Echelon Approach. *International Journal of Remote Sensing* 21, 2301-2306.

Smits, P and W. Myers. 2000. Echelon Approach to Characterize and Understand Spatial Structures of Change in Multitemporal Remote Sensing Imagery. *IEEE Transactions on Geoscience and Remote Sensing*, 38(5):2299-2309.

Appendix A. Public Packages for Portraying Polypatterns

Since our reference is to landscapes, pursuing patterns purely in the abstract would be partially paradoxical. Software for preparing and portraying images like those shown earlier thus becomes pertinent. Programming for polypatterns and is an evolutionary endeavor. A modular system in generic C language (including source code) is available on the Internet for downloading under the acronym PSIMAPP {Progressively Segmented Image Modeling As Polypatterns) for which usage is outlined in Appendix B. Actual presentation of graphics for viewing, however, is a specialized sector of software with system specificity and device dependence. As an expedient, we patronize other packages for this purpose.

The primary place of polypatterns is at the interface of image analysis and geographical information systems (GIS) that manipulate map data. Since GIS has widespread usage, commercial compatibility from polypattern processing has been directed toward the ArcGIS family of facilities by Environmental Systems Research Institute (ESRI) which has a majority of the American market (see www.esri.com). However, it will also be helpful for learning purposes if we delve a little deeper into imaging via some software for PC-compatible and McIntosh computers called MultiSpec that NASA has sponsored through Purdue University in the interest of public outreach and education.

A.1 MultiSpec for Multiband Images and Ordered Overtones

The URL for the MultiSpec website is not a simple one to remember, but it is easily located by doing a web search for MultiSpec. There is no charge for MultiSpec, and it has the added advantage of not requiring administrative installation on the computer. Simply (double) clicking the icon in a folder is sufficient to launch the application. MultiSpec is also more tolerant than most imaging software when its preferred file of auxiliary information about the image is absent. If the prospective user knows the nature of the image information, MultiSpec will allow the particulars to be entered explicitly in dialog boxes. However, MultiSpec does have some peculiarities of terminology regarding images that come from early remote sensing systems and can trip the uninitiated. The first of these is that it uses *channels* instead of *bands* in referring to sets of signals.

Fig. A.1 shows the initial desktop view for MultiSpec when it is launched. We recommend that the text box be resized by dragging a corner, and then moved to the right side by dragging the banner as is shown in Fig. A.2 so that it is not concealed beneath an image that is about to be loaded.

To begin the process of opening an image, one can use either the File > Open Image menu or the folder icon at the left side below the menus. For purposes of illustration, we will first consider loading band 3 of the low resolution Landsat MSS multi-band image as shown in Fig. 1.2 of the first chapter. Thereafter, loading of the corresponding polypattern ordered overtones from this image will be illustrated.

Before proceeding with these illustrations, however, it is essential that we clarify the nature of gray-tone renderings versus color renderings. A gray-tone image is a special case of a color image when working with an

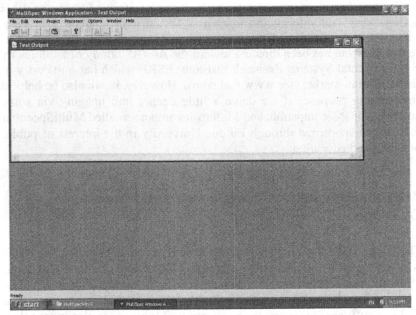

Fig. A.1 Initial screen for MultiSpec after launching and maximizing.

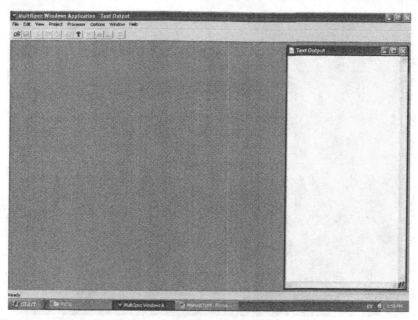

Fig. A.2 MultiSpec desktop after resizing Text Output box and moving it to the right.

RGB display facility that is capable of showing color. In this case, the gray-tone is created from three identical signal bands with one being sent to each of the Red, Green, and Blue components. Most such display facilities offer a choice of single-band mode or color mode. In single-band mode, the user selects one of the available bands with the facility taking care of making a triplicate version. It is also possible to obtain the same result by using color mode and specifying the same band for all colors.

At this juncture of our exposition, we will be specifying all of the information about the image by explicit entries in the dialog boxes of MultiSpec. This information is obtained either by accessing some sort of 'header' file, or from printed documentation accompanying the image. For our example of displaying an individual band of a multi-band image, the pertinent specifications are 1000 rows of pixels, 700 columns of pixels, and 5 signal bands that are interleaved by pixel with one-byte binary encoding. The infrared band that we propose to display is the third band.

By going to File > Open image in the menu, we obtain the initial dialog box shown in Fig. A.3 whereby we navigate to the directory containing the image data. However, here is where we encounter the first aspect of MultiSpec that is somewhat unusual; which is that MultiSpec does not automatically recognize a data file with a .BIP extension as containing

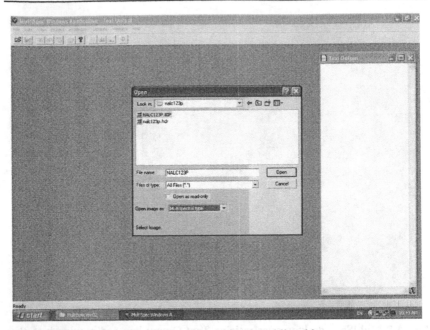

Fig. A.3 Multispec first dialog box for opening a multiband image.

image data. Instead, MultiSpec expects that the band-interleaved-by-pixel arrangement will have a .BIS extension that stands for Band Interleaved by Sample. In other words, MultiSpec considers a pixel to contain a sample of signals coming from the area that is encompassed by the pixel. It is simple enough to resolve this discrepancy by using the dialog box to request that all files in the folder be listed so that we can make the selection of the .BIP file that contains the image data. Since this is multiband image data, we also need to specify that the image be opened in Multispectral mode.

Having made these initial selections, we can activate the *Open* button to move on to the next dialog box as shown in Fig. A.4. At this juncture we specify the number of rows of pixels (1000), the number of columns of pixels (700), the number of bands = channels (5), and the format of the data as one-byte per band and BIS (according to MultiSpec terminology). Since there are no additional bytes in this data file, we can proceed with the *OK* button.

We see that there are more specifications to be made, since we are presented with another dialog box as shown in Fig. A.5. Here is where we indicate that we wish to see a single one of the bands in gray-scale mode instead of using RGB color mode, and specify which one of the bands

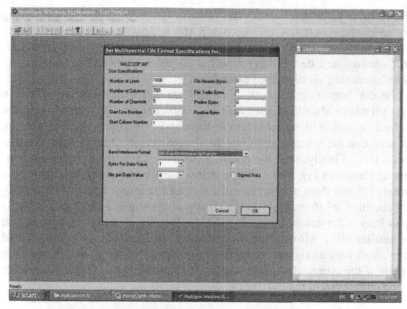

Fig. A. 4 MultiSpec second dialog box for opening an image.

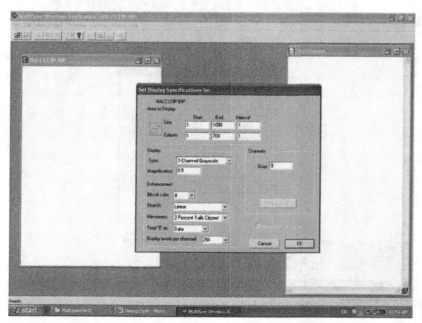

Fig. A. 5 MultiSpec third dialog box for opening a multiband image.

(channels) is to be displayed. The remaining specifications in this dialog box offer the possibility of improving the appearance of images having low contrast. As a first look, however, it is appropriate to accept the default settings and decide later whether to try improving the contrast by more 'stretching' as discussed elsewhere.

The *OK* button again takes us to still another dialog box as shown in Fig. A.6 where MultiSpec asks for permission to calculate a new histogram of the distribution of signal values in the bands. We grant this permission, but note that the folder containing the image data must have write permission active. Finally, then we obtain a display of the image in a new window as shown in Fig. A.7. This window can be expanded by dragging a corner, but that does not increase the size of the image itself. The facility for 'zooming in' to magnify is the button at the top having the larger hills as an icon. Conversely, 'zooming out' is done by clicking the button with the smaller hills. Moving the image around or 'panning' is accomplished with scroll bars along the margins of the image window, which are only present if the size of the image window is insufficient to display the entire image at the current magnification.

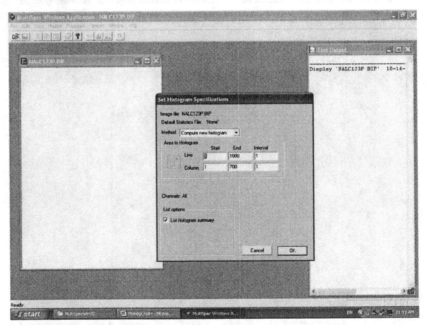

Fig. A.6 MultiSpec fourth dialog box for opening an image.

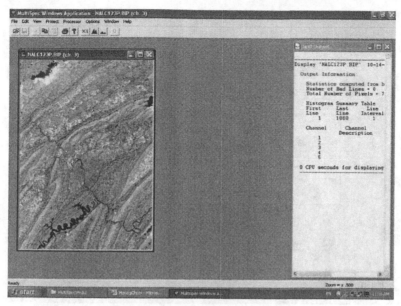

Fig. A.7 Gray-scale display of channel (band) 3 in MultiSpec.

Having used MultiSpec to view a band from a conventional multi-band image data file, we now use it for preliminary viewing of polypattern ordered overtones. Prior to doing so, however, the basic polypattern processing must be conducted. Appendix B contains an introduction to a set of PSIMAPP software modules for polypattern processing. This viewing would follow the basic pattern processing operation covered in Appendix B-1, and the present illustration assumes that NALC123m is used as the name for the resulting polypatterns.

We again initiate the open-image sequence of MultiSpec, but this time we choose the file having a .BSQ extension and we accept the *default* image mode as shown in Fig. A.8. Clicking the *Open* button would then bring up the second dialog box as shown in Fig. A.9.

Things to note in the second dialog box for opening the overtones of A-level patterns are that the entire range of signal values has been specified for display and that zero is being treated as background for the image. Clicking the *OK* box will then directly activate the image window as shown in Fig. A.10, after which the image window can be expanded by zooming and panning as described earlier.

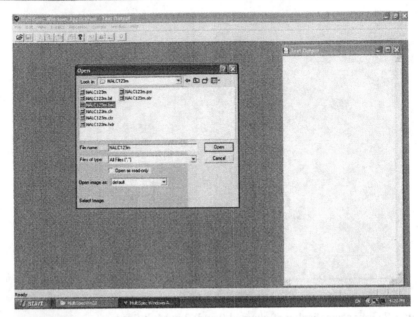

Fig. A.8 Illustration of first MultiSpec dialog box for polypattern overtones.

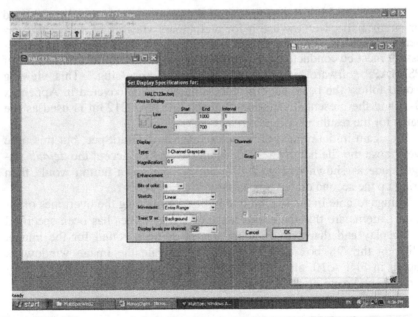

Fig. A.9 MultiSpec second dialog box for opening polypattern ordered overtones.

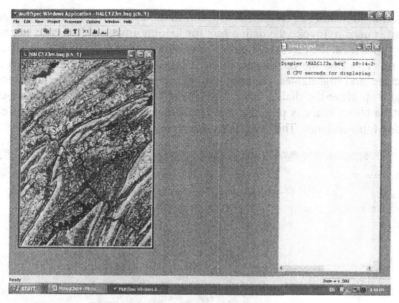

Fig. A.10 MultiSpec display of polypattern ordered overtones.

A.2 ArcExplorer

ESRI makes publicly available through its website (www.esri.com) an image and map viewer called *ArcExplorer* that accommodates spatial data products from the *ArcGIS* family. This viewer requires installation under administrative privileges on desktop Microsoft Windows PC-compatible computing systems. ArcExplorer requires that information about the image be contained in a textual file having .HDR extension, which is automatically prepared by the PSIMAPP software as described in Appendix B. An example of the contents of this *header* file for the illustrative data is as follows.

```
BYTEORDER    M
LAYOUT       BSQ
NROWS        1000
NCOLS        700
NBANDS       1
NBITS        8
BANDROWBYTES      700
TOTALROWBYTES     700
BANDGAPBTYES      0
```

After launching ArcExplorer, the icon in the upper row with a large plus sign is used to invoke the dialog box for loading images. It may be necessary to refresh the directory tree with the button provided in this dialog box in order to find the polypattern file (.BSQ extension) of interest. Having located and highlighted the appropriate file click the Add Theme button and then close the dialog box. The image will not actually be displayed until a check mark is placed in the box beside its identifier along the left side of the desktop. The result is as shown in Fig. A.11.

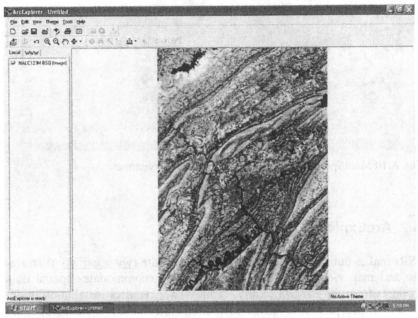

Fig. A.11 ArcExplorer display of polypattern overtones like Fig. A.10.

Appendix B. PSIMAPP Software

Appendix B introduces modular software in C-language for the α-scenario of polypattern processing written by Wayne Myers. This set of software carries the acronym PSIMAPP for Progressively Segmented Image Models As PolyPatterns. PSIMAPP is a collection of software modules written in generic C language, with each module addressing particular polypattern processing procedures. The C code in a module does not target any particular computer platform, and specifications are read in text form from a so-called RUN file using certain protocols. The bulk of usage, however, has been on PC-compatible desktop computers running Microsoft Windows operating system. To give PSIMAPP the look and feel of Windows-oriented dialog boxes, a Visual-C front-end has been developed by Ningning Kong. This introduction is set forth entirely in terms of the front-end environment which takes input specifications from the user in a Windows fashion from which it prepares an appropriate RUN file that it uses to execute the respective module. Appendix C provides a brief introduction to direct development of RUN files using a text editor for purposes of processing on other platforms. Instead of giving a specific Internet URL, we simply advise the user to do a web search for PSIMAPP. The forms encountered on a downloaded version may not be exactly like those shown here since the PSIMAPP modules and interface undergo evolutionary developmental changes.

PSIMAPP does not require administrative installation on the computer, and is launched by simply double-clicking on a globe icon. The PSIMAPP initial dialog box gives a list of modules that are available in this mode, offers access to brief explanatory help files, and requires the user to navigate to and select a so-called *path* file having .PTH extension that is contained in the PSI5 folder where the modules reside. The purpose of this selection is to enable paths for executing modules to be established automatically for the remainder of the session. This initial screen is shown in Figure B.1. The user then selects a module from the list on the left-hand side and clicks the *Continue* button in order to proceed with input and execution for that module.

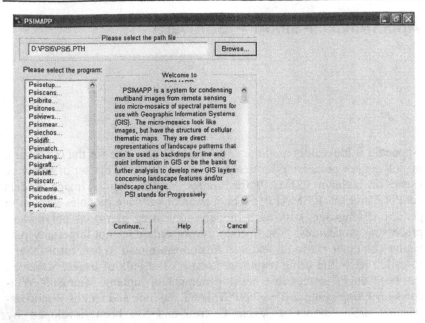

Fig. B.1 Opening dialog box for PSIMAPP interface to software modules.

The remainder of this Appendix is divided into numbered sections that address usage of particular modules as follows:

B.1.   Polypatterns from Pixels.

B.2.   Supplementary Statistics

B.3.   Collective Contrast.

B.4.   Tonal Transfer Tables.

B.5.   Combinatorial Contrast.

B.6.   Regional Restoration.

B.7.   Relative Residuals.

B.8.   Direct Differences.

B.9.   Detecting Changes from Perturbed Patterns.

B.10. Edge Expression.

B.11. Covariance Characteristics.

B.1 Polypatterns from Pixels

Extraction of polypatterns from multi-band pixel data according to the □-scenario is conducted by the PSISCANS module. The multi-band pixel data file is expected to have BIP (band interleaved by pixel) format with byte-binary encoding of signal values, and the user must know the numbers of row, columns, and bands. Creation of the polypatterns shown in Figures A.10 and A.11 will serve here for purposes of illustration. This dataset has 1000 rows, 700 columns, and 5 bands. However, the fifth band will be excluded because it is a synthetic band that was derived from the other bands.

The first PSIMAPP dialog box for PSISCANS is shown in Figure B.1.1 after specifications have been entered. The output files will have NALC123m as the base name. The defaults of 2500 B-level patterns in 12 scans have been accepted.

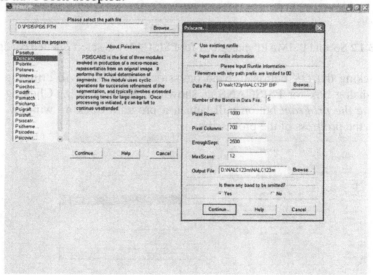

Fig. B.1.1 First PSIMAPP dialog box for PSISCANS module.

Pressing the *Continue* button brings up the second dialog box for selecting bands to be omitted as shown in Figure B.1.2. Band 5 has been unchecked to indicate omission.

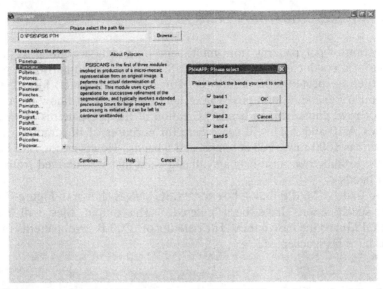

Fig. B.1.2 Second PSIMAPP dialog box for PSISCANS to select omitted bands.

Clicking the *OK* button in band omission dialog box will bring up the final dialog box for PSISCANS as shown in Figure B.1.3. Clicking the *Execute the program* button will launch a program tracking window that shows the progress of the processing.

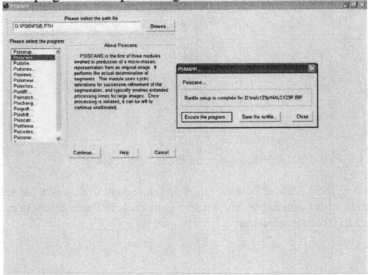

Fig. B.1.3 Final PSIMAPP dialog box for PSISCANS before executing module.

B.2 Supplementary Statistics

There are two tables of supplementary statistics for A-level patterns that pertain to categorical classification for thematic mapping of landscape components. One of these files has a .MEN extension and the other has a .VAC extension. Considering the contents of these tables is not the current concern, but it is an important issue that their computation requires the original image data as an input. Therefore it is prudent to produce them promptly upon obtaining the polypatterns so that they are part of the patterns package. This is accomplished by running the PSISCATR module, and the foregoing illustration is continued accordingly with Figure B.2.1 showing the first dialog box.

For the PSI Name entry, navigate to the polypattern folder and choose the file with the .PSI extension. The data file is the same one that was used as input to the PSISCANS module. The record size is the number of signal bands that are recorded in the image data file, regardless of whether they were all used for procuring polypatterns. The *Chatter* and *Incremental* entries can generally be allowed to default. Any signal bands that were omitted in procuring polypatterns must also be omitted for this purpose. Clicking *Continue* carries on to the next dialog box, which depends on whether items are to be omitted. The dialog box for omitting items is shown in Figure B.2.2. If there were no items to be omitted, then the final dialog box would appear at this point as shown in Figure B.2.3. Then clicking the *Execute the program* button initiates the processing.

B.3 Collective Contrast

Beyond simple viewing of the PSISCANS output as a gray-scale image of ordered overtones, it is necessary to compile a pair of relative brightness tables, one of which is in a file with .BRS extension and the other in a file with .BRI extension. The contents of these tables are described in Chapter 3. The present concern is only with preparation of the tables using the PSIBRITE module of the PSIMAPP system. The opening dialog box for PSIBRITE from PSIMAPP is shown in Figure B.3.1.

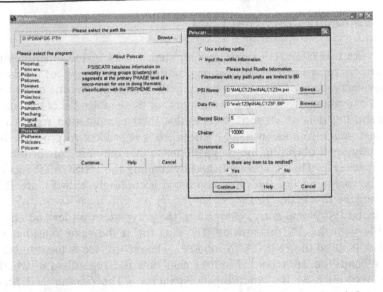

Fig. B.2.1 First dialog box of PSISCATR module for supplemental statistics.

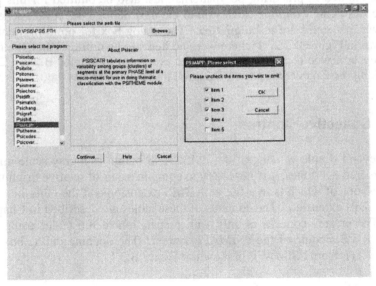

Fig. B.2.2 Second PSISCATR dialog box for omitting signal items (bands).

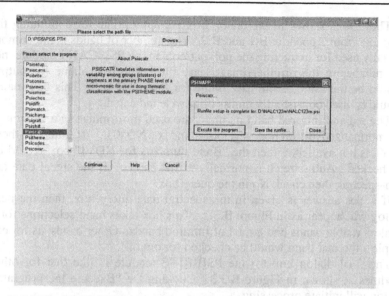

Fig. B.2.3 Final PSISCATR dialog box for tables of supplemental statistics.

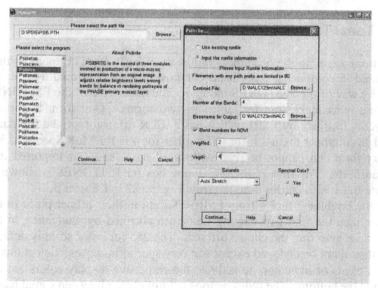

Fig. B.3.1. Opening PSIMAPP dialog box for PSIBRITE module with selections.

The entries shown in the dialog box are appropriate for continuing the example from Appendix B.1 and B.2. The number of bands is the number actually used for extracting the polypatterns in the PSISCANS module, not the number in the original image data file. The base name for output should be the same as the base name for the .CTR file. This base name should be used consistently unless otherwise indicated.

The red and infrared band numbers are used in formulating a version of the normalized difference vegetation index (NDVI). If either of these bands is not available, then the "Band numbers for NDVI" box should be unchecked. Auto-stretch is normally a first choice. If the signal data are non-spectral, then check No in the query box.

If a Yes answer is given in the spectral data query box, then the next dialog will appear as in Figure B.3.2. This box takes band selections for a total of visible bands and a total of infrared bands. Other bands, as for example a thermal band would be checked for omit.

The final dialog box for the PSIBRITE module is like that for other modules as shown in Figure B.3.3. Clicking the "Execute the program" button will initiate processing.

B.4 Tonal Transfer Tables

After compiling the brightness tables with the PSIBRITE module, they are used in composing three-band composite images as tonal transfer tables by way of the PSITONES module. The major role of the PSITONES module is to make transfer tables in two forms. One form is for the ArcXXX facilities by ESRI, which has a .CLR extension on the file name and has columnar textual form. The other form is for MultiSpec, and has binary form that cannot be read with word processors nor imported into spreadsheets. The first PSIMAPP dialog box for PSITONES is shown in Figure B.4.1 with entries used to create Figure 3.2 in Chapter 3.

If the brightness tables from PSIBRITE are in their proper place in the folder, then the first two naming items are obtained by navigating to the folder and selecting the choice offered. The 255 for color scale is default, and should not be changed except for very special purposes. Selections of signal bands or indicators to activate the respective display colors are entered at the bottom. The radio buttons control choice between signal bands from the .BRS brightness table or integrative indicators from the .BRI table. Using identical entries from the same brightness table for all three will give a gray-tone image. The selection shown in the example is the vegetation indicator, which is the sixth entry in the .BRI table.

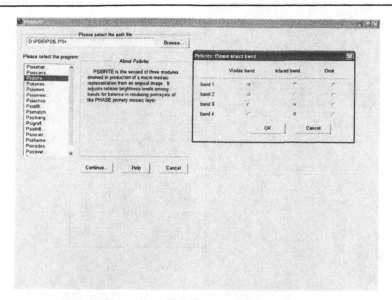

Fig. B.3.2 Second PSIMAPP dialog box for PSIBRITE module with spectral data.

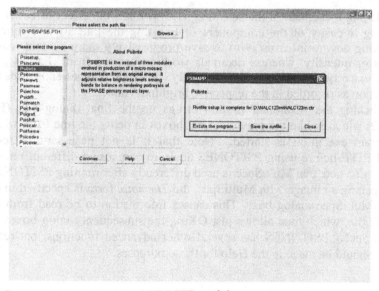

Fig. B.3.3 Final dialog box for PSIBRITE module.

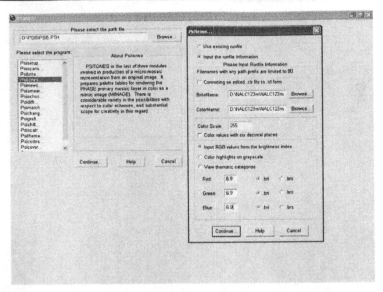

Fig. B.4.1 First PSIMAPP dialog box for PSITONES module.

Placing a decimal on the selection as shown has the effect of either reducing intensity of the component or boosting the intensity. Decimals working downward from .9 to .6 give progressively more evident reductions of intensity, whereas decimals working upward from .1 to .4 give progressively stronger boosts of intensity. A simple integer entry uses the selection as recorded in the respective brightness table.

Clicking the "Continue" button brings up the final dialog box (Fig. B.4.2) which is like the previous final boxes in being the one from which program execution is started. Note that it is not necessary to rerun PSIBRITE before using PSITONES again to compose a different rendering. Also note that MultiSpec is used differently after running PSITONES. In opening an image with MultiSpec, the *Thematic* form is specified in the first MultiSpec dialog box. This causes information to be read from the .TRL file, which then allows just OKing the subsequent dialog boxes for MultiSpec. PSITONES has several other advanced functions, but reference should be made to the Help for those purposes.

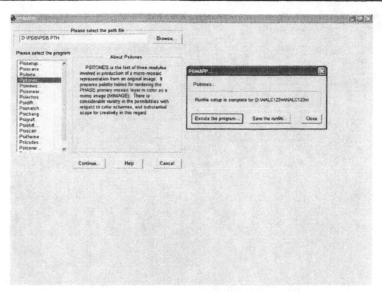

Fig. B.4.2 Second PSIMAPP dialog box for PSITONES module.

B.5 Combinatorial Contrast

There are rather large numbers of possible choices for triplets in using the PSITONES module, and it may not be evident which of those are potentially more interesting than others. The PSIVIEWS module provides some suggestions of combinations that will have strong contrast, but it does not specify which item of the three items should be placed on which color. That is left to the creative investigation of the user.

The output of PSIVIEWS is a short textual file having a .VUE extension that can be opened in a text editor or word processor. Each line has a triplet of numbers. A negative sign on the number indicates that the item is in the .BRS file. Absence of a sign indicates that the item is in the .BRI file. The dialog for this is quite simple, with the first box as shown in Figure B.5.1 serving to select the source name with .PSI extension. The same name is used with a .VUE extension for output. The second (last) dialog box is like the ones for modules described above.

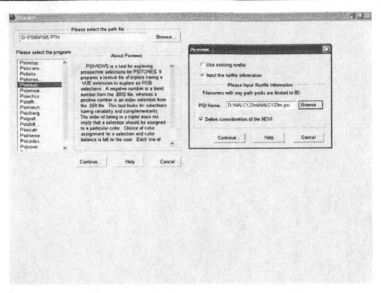

Fig. B.5.1 First PSIMAPP dialog box for PSIVIEWS module.

B.6 Regional Restoration

The PSIECHOS module provides for approximate restoration of multi-band image data files from B-level patterns. Restoration can be limited to a selected block of rows and columns of pixels as well as to particular bands. There are also options for contrast enhancement during restoration. The first PSIMAPP dialog box for PSIECHOS is shown in figure B.6.1 with entries for the image data considered above.

The dialog box entries are the defaults or most general ones. Specifying zero for all the starts and ends will restore the entire image, and the checked box restores all bands. Trim 2 gives a full range linear contrast stretch with no saturation. The *Chatter* value just controls how often progress is reported to the screen during execution. With these general settings, the *Continue* button transfers directly to the last dialog box where execution is initiated. There are additional options as mentioned above that would invoke additional dialog boxes, for which the PSIMAPP *Help* information should be consulted.

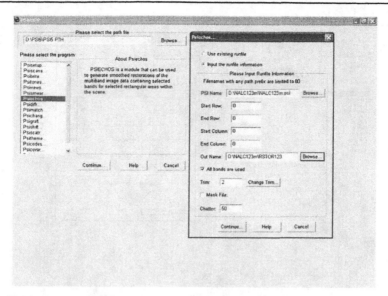

Fig. B.6.1 First PSIMAPP dialog box for PSIECHOS module.

B.7 Relative Residuals

The PSISMEAR module generates information needed to make an image showing the differences in degree of approximation when restoring multi-band image data from B-level patterns. The residuals are measured by length of the difference vector between the signal values of the actual pixel and the approximated pixel. The vector lengths are scaled into 255 classes, with the classes being numbered 1 through 255 and 0 indicating missing data. The mean vector length and pixel count for each class are reported in a special version of the .BRS brightness table having the same root name as the residual image data. The first column is residual class number, the second column is average length for the residual vectors in the class, and third column is number of pixels in the class. This is not an actual brightness table, and must not be used for setting up the image display in PSITONES. Intensity scales for setting up an image in PSITONES are contained in the .BRI brightness table. All PSITONES selections must be made from the .BRI file. Figure B.7.1 shows the first PSIMAPP dialog box for PSISMEAR.

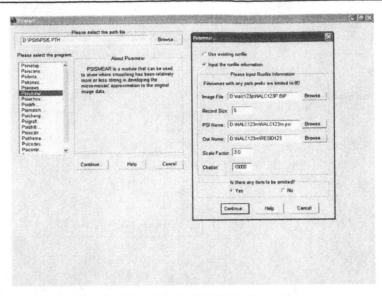

Fig. B.7.1 First PSIMAPP dialog box for PSISMEAR module.

Both the original image and the polypatterns are required as input to this module. The record size refers to the original image, and any bands that were omitted must be omitted again. Making the scale factor smaller will make smaller residuals more evident in an image, whereas making the scale factor larger will tend to suppress smaller residuals.

If bands are to be omitted, there is a subsequent dialog box for indicating which bands to retain. Otherwise, the *Continue* button invokes the last dialog box where execution is initiated.

After running PSISMEAR, the PSITONES module must be run to set up the image of residuals. DO NOT run the PSIBRITE module prior to PSITONES. The PSITONES selection is ONLY made from BRI. The number 1 selection is an upward scaling of residual classes. The number 2 selection is a downward scaling of residual classes. The number 3 selection is a dummy for nullifying a color. Use 1 1 1 for RGB selections in PSITONES to get a gray-scale image of residuals with largest residuals being brightest. Use 2 2 2 for RGB selections in PSITONES to get a gray-scale image of residuals with largest residuals being darkest. A selection of 1 3 2 is suggested for a color rendition of the residual image in which smaller residuals are bluish and larger ones reddish with intermediates being shades of magenta.

B.8 Direct Differences

The PSIDIFFR module provides for determining change vectors from A-levels of two spatially matched polypatterns having the same signal bands but for different occasions. The images to be compared for change must have the same directional orientation and the same pixel resolution (size), but they need not have the same area of coverage. Coupling of the two images is accomplished by specifying row and column positions in each that mark the same location. The coverage of the output grid is the same as that of the first image, with all non-overlapping positions being given a zero value. *It is important to note* that the output does not include a header (.HDR) file for the grid. A copy of the header file for the first image must be renamed and modified as necessary to suit the output grid before a viewer can be invoked. Figure B.8.1 shows the first PSIMAPP dialog box for this module.

In this dialog box, PHASE is an older term used as a designation for the A-level of patterns. The first occasion is usually treated as Grid A and the second as Grid B, although it is permissible to do the reverse. Provision is made for designating lists of patterns that are not to be considered as being change. These are called *hidden clusters* in the dialog box. If there are no lists of hidden items, then uncheck the corresponding boxes.

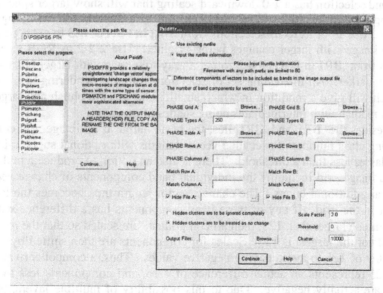

Fig. B.8.1 First PSIMAPP dialog box for PSIDIFFR direct change vector module.

The output of this module has parallels to that of the PSISMEAR module for determining residuals as described in Appendix B.7. The computed change vector lengths are scaled into 255 classes, with the classes being numbered 1 through 255 and 0 indicating missing data. The mean vector length and pixel count for each class are reported in special version of a .BRS brightness file. Note carefully that this version of the .BRS file is not intended for selection in a PSITONES rendering of the resulting difference image. This special version of the .BRS table has three columns. The first column is class number, the second is average change vector length for the class, and the third is number of pixels belonging to the class. The class intervals in the .BRS table are controlled by a scale factor. The scale factor is a multiplier for the mean change vector length. This multiple of the overall mean length is scaled equally into the first 254 classes, and anything larger appears in class 255. Making the scale factor smaller will make smaller changes more evident, whereas making the scale factor larger will tend to suppress smaller changes. There is also an option to set a threshold for class number such that vector lengths in smaller classes will not be indicated as change in a viewer.

A special version of a .BRI brightness table is produced for use in rendering the change image via PSITONES. The first selection has a 0-1 upward scaling that will show larger spectral changes in brighter tones. The second selection has a 1-0 downward scaling that will show larger spectral changes in darker tones. The third selection is a dummy for nullifying a color. Use 1 1 1 as RGB selections from .BRI in PSITONES to get a graytone image with larger changes being brighter. Use 2 2 2 as RGB selections from .BRI in PSITONES to get a gray-tone map with larger changes being darker. An interesting set of selections for color rendering is 1 3 2, which will give a rendition in which smaller changes are bluish and larger ones reddish with intermediates being shades of magenta.

Besides the .BRS and .BRI files, the basic output is an image grid containing class numbers for change vector lengths with optional suppression of classes less than a threshold. There is also an option to produce a multiband image grid including the individual band components of change vectors as additional layers. The change vector length then becomes the last layer in the grid. If any of these band components has a difference with magnitude larger than 125, then all components are scaled so that the magnitude of the largest is 125. Scales for components are then shifted by the amount of 125 to avoid having negative values. Thus, a component value of 125 represents an actual difference of zero, and components less than 125 are actually negative. Due to this possibility of multiple layers, the output grid file automatically has a .BIP extension on the name instead of a .BSQ extension. Thus, the .BIP file must be selected in a viewer.

B.9. Detecting Changes from Perturbed Patterns

Two modules are involved in detecting changes from perturbation of arrangements for counterpart patterns in two instances of imaging. The first module is PSIMATCH which determines counterpart patterns and thus establishes a linkage between occasions. The second is PSICHANG which works much like PSIDIFFR, except that it uses the linkage instead of direct differences.

The images to be compared for change must have the same directional orientation and the same pixel resolution (size), but they need not have the same area of coverage. Coupling of the two images is accomplished by specifying row and column positions in each that mark the same location. The coverage of the output grid is the same as that of the first image, with all non-overlapping positions being given a zero value.

Figure B.9.1 shows the first PSIMAPP dialog box for the PSIMATCH module. In this dialog box, PHASE is used as a designation for the A-level of patterns.

The first occasion is usually treated as the base and the second occasion as the linkage, although it is permissible to do the reverse. Provision is made for designating lists of patterns that are not to be considered as being change. These lists are specified in *hide files*. If there are no lists of hidden items, then uncheck the corresponding boxes. Output is a linkage file

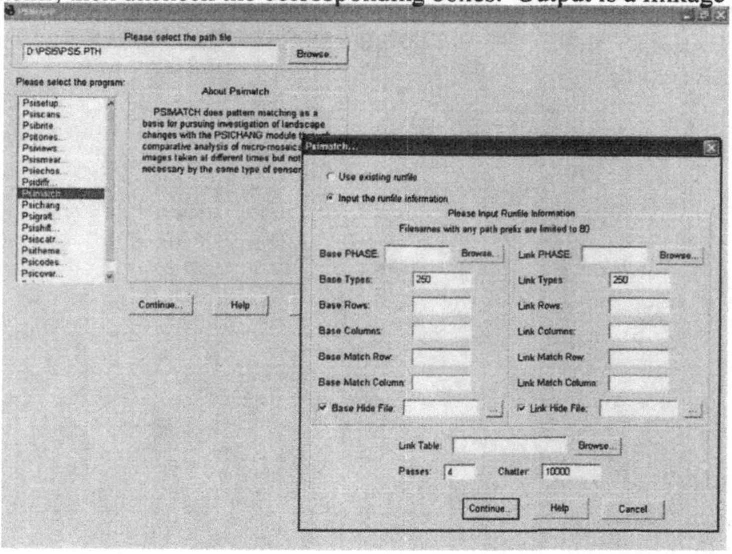

Fig. B.9.1 First PSIMATCH dialog box for PSIMATCH module.

that gives counterpart patterns for the occasions. The main part of the name for the linkage file must be specified, but PSIMAPP supplies a .TBL extension.

Figure B.9.2 shows the first PSIMAPP dialog box for the PSICHANG module that uses the output of the PSIMATCH module. This module parallels the PSIDIFFR module. The occasion used as Base in PSIMATCH is used as Grid A; however, the table for this grid is the linkage table from PSIMATCH instead of a .CTR file. *It is important to note* that the output does not include a header (.HDR) file for the grid. A copy of the header file for the first image must be renamed and modified as necessary to suit the output grid before a viewer can be invoked.

The computed change vector lengths are scaled into 255 classes, with the classes being numbered 1 through 255 and 0 indicating missing data. The mean vector length and pixel count for each class are reported in a special version of a .BRS brightness file. Note carefully that this version of the .BRS file is not intended for selection in a PSITONES rendering of the resulting difference image. This special version of the .BRS table has three columns. The first column is class number, the second is average change vector length for the class, and the third is number of pixels belonging to the class. The class intervals in the .BRS table are controlled by a scale factor. The scale factor is a multiplier for the mean change vector length. This multiple of the overall mean length is scaled equally into the

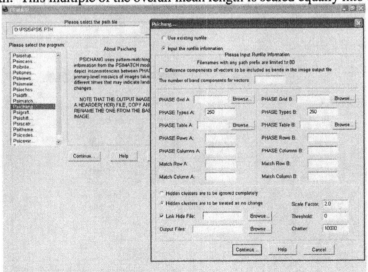

Fig. B.9.2. First PSIMAPP dialog box for PSICHANG module.

first 254 classes, and anything larger appears in class 255. Making the scale factor smaller will make smaller changes more evident, whereas making the scale factor larger will tend to suppress smaller changes. There is also an option to set a threshold for class number such that vector lengths in smaller classes will not be indicated as change in a viewer.

A special version of a .BRI brightness table is produced for use in rendering the change image via PSITONES. The first selection has a 0-1 upward scaling that will show larger spectral changes in brighter tones. The second selection has a 1-0 downward scaling that will show larger spectral changes in darker tones. The third selection is a dummy for nullifying a color. Use 1 1 1 as RGB selections from .BRI in PSITONES to get a gray-tone image with larger changes being brighter. Use 2 2 2 as RGB selections from .BRI in PSITONES to get a gray-tone map with larger changes being darker. An interesting set of selections for color rendering is 1 3 2, which will give a rendition in which smaller changes are bluish and larger ones reddish with intermediates being shades of magenta.

Besides the .BRS and .BRI files, the basic output is an image grid containing class numbers for change vector lengths with optional suppression of classes less than a threshold. There is also an option to produce a multiband image grid including the individual band components of change vectors as additional layers. The change vector length then becomes the last layer in the grid. If any of these band components has a difference with magnitude larger than 125, then all components are scaled so that the magnitude of the largest is 125. Scales for components are then shifted by the amount of 125 to avoid having negative values. Thus, a component value of 125 represents an actual difference of zero, and components less than 125 are actually negative. Due to this possibility of multiple layers, the output grid file automatically has a .BIP extension on the name instead of a .BSQ extension. Thus, the .BIP file must be selected in a viewer

B.10 Edge Expression

The PSIEDGES module determines edge relations among A-level polypatterns that are of interest for analysis of context. The question is to what extent some patterns tend to occur more often as neighbors than others beyond a random expectation.

There are two output files. One is a textual file having an .EDG extension that contains an edge 'profile' for each A-level pattern. The other is a full file of edging data in byte binary mode. In the .EDG file each pattern has a line of profile information. The first field on this line is pattern number to which the profile pertains. The second field is frequency order for the pattern. The third field is edge order for the pattern, excluding self-edges and outer edges. The fourth field is a self-edge fraction. The remaining nine fields are minimum number of other edge segments to attain deciles of edges, excluding self-edges. Detail lines following the profile are optional. The detail lines give information on sharing of edges with particular patterns. An edge pattern is reported as a decimal number. The whole number part is the number of the edge pattern. The fractional part is a 100-step index of departure from expectation for a random grid having the same frequencies for the respective patterns. The sign of the number indicates direction of departure from random expectation. It is the higher-order patterns in this regard that will be listed, where higher-order refers to frequency of occurrence as an edge type. The ordering is thus according to decreasing domination of edge patterns.

The second output file has an .EGN extension and contains complete information on sharing of edges in a binary format. There are as many bytes for each pattern as there are patterns (i.e., 250). Each byte encodes edging for the corresponding pattern number. Zero indicates correspondence to random expectation. Numbers in the range 1-100 indicate increasing negative departure from expectation. Numbers in the range 101-200 indicate increasingly positive departure from expectation.

The grid is processed in vertical strips. The size of the strip determines the amount of high speed memory needed. If the module aborts due to insufficient memory, then reduce the strip size. Figure B.10.1 illustrates the first PSIMAPP dialog box for the module. The number of *edging lines* determines how many edge neighbor patterns can be listed along with the profile. Activating the *Continue* button brings up the last dialog from which execution is initiated.

B.11 Covariance Characteristics

Covariance, or joint variability, between different bands is of statistical interest for various purposes. It is the basis for several statistical classifiers, transformations, and principal components. The covariance values among bands are typically arranged in the form of a matrix having band variances on the principal diagonal. Band standard deviations are the square roots of the variances. A covariance matrix can be converted to a correlation matrix by dividing each covariance by the product of the standard deviations for the pair of variables involved. A correlation matrix always has 1.0 as the value of all diagonal elements.

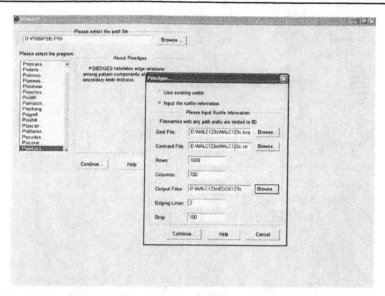

Fig. B.10.1 First PSIMAPP dialog box for PSIEDGES module.

The PSICOVAR module calculates an overall covariance matrix and the covariance matrix for each A-level pattern. Both the original image data file and the polypattern files are required for these computations, and the PSISCATR module must be run prior to running PSICOVAR. Figure B.11.1 shows the first **PSIMAPP** dialog box.

The overall covariance matrix takes the form of a textual file having a .COV extension on its name, with all elements of the matrix present in the file in row-wise order with five elements per line separated by spaces. The covariance matrices for individual A-level patterns appear in upper triangular form in another textual file having a .CVS extension. Each of the matrices is introduced by a line containing just the word CLUSTER followed by the A-level pattern number. The remaining lines for that matrix each contain five of the upper triangular elements.

If signal variables (bands) have been omitted in forming the polypatterns, then they must also be omitted for this module. A dialog box is available for omitting items.

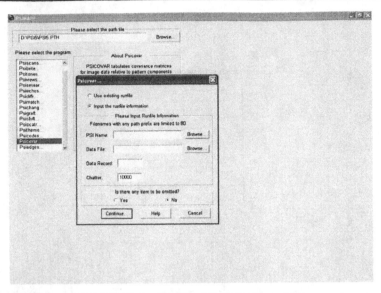

Fig. B.11.1 First PSIMAPP dialog box for PSICOVAR module.

Appendix C. Details of Directives for PSI MAPP Modules

The PSIMAPP modules actually use a textual rather than a graphic interface. The specifications for operating a module are entered into a textual file that usually carries the ending (extension) .RUN on the file name. The text in this file must be simple ASCII characters with no auxiliary formatting information, not even tabs. Therefore, one should not use sophisticated word processing facilities in preparing these specification files unless the output is stored as simple text to strip internal formatting codes. Each line of the file should end with carriage return. These files are structured so that annotation regarding the specifications can be incorporated directly in the file. To make this possible, any line that has an exclamation mark in the first column is ignored entirely by the facility that does the reading of specifications. The actual specifications are substituted into lines having the form:

$$\{specname=specvalue\} \hspace{3cm} (C-1.1)$$

where the 'specvalue' placeholder is replaced by the value of the specification. No blank spaces are permitted within the curly brackets. For instance, the PSISCANS file of directives (without annotation) for the polypatterns shown in Figures A.10 and A.11 is:

```
{DataFile=D:\nalc123p\NALC123P.BIP}
{DataBands=-5}
{UseBands=D:\NALC123s\usebands.txt}
{PixelRows=1000}
{PixelCols=700}
{OutputName=D:\NALC123s\NALC123s}
{EnoughSegs=2500}
{MaxScans=12}
[RUN] {DataFile} {DataBands} {UseBands} {PixelRows} {PixelCols}
[RUN] {OutputName} {EnoughSegs} {MaxScans}
```

This set of specifications was actually set up by the PSIIMAP interactive front end. A fully annotated version that would typically be used apart from the front end is a follows.

```
!   DataFile is the name of the multiband image data file that is to be
!   converted.  The data must be byte-binary, and must be arranged in
!   band-interleaved-by-pixel (BIP) sequence.  The root file name must be
!   limited to 8 characters, and extension (if any) must be specified.
!   If the file is not in the default directory, then the path must also
!   be specified.  Blank spaces are not allowed in names comprising the
!   path.
    {DataFile=D:\nalc123p\NALC123P.BIP}
!
!   DataBands is the number of bands recorded for each pixel in the DataFile,
!   including any that are to be ignored during conversion.
    {DataBands=-5}
!
!   UseBands is the full file specification for supplemental information on
!   band selection if processing is limited with respect to bands.  If
!   processing is limited with respect to bands, then the number of DataBands
!   as given above must also be made negative -- otherwise, the UseBands
!   specification is treated as just a placeholder.  A supplemental file
!   must be simple text (not word processor document) with spaces serving
!   as delimiters (do not use tabs).  In this supplemental file, list the
!   sequence numbers of the bands to be used.  For bands that are to be
!   included in the segmentation, but not appear in the .ctr file, put a
!    plus sign on the band number.  Otherwise, do not use plus signs.
!   Negative signs are used to indicate ranges of band numbers.  List the
!   starting number in the band range as usual (with or without a plus sign
!   as appropriate), then leave a space and give the ending band number in
!   the range with a negative sign.  All bands in the range will be treated
!   the same with respect to inclusion in the .ctr file.  Once again, if all
!   bands are to be used in all aspects of conversion, then nothing needs
!   to be done here.
    {UseBands=D:\NALC123s\usebands.txt}
!
!   PixelRows is the number of rows of pixels in the image data file.
    {PixelRows=1000}
!
!   PixelCols is the number of columns of pixels in the image data file.
    {PixelCols=700}
!
!   OutputName is the root file name (with path if needed, but no extension)
!   for the output files.  Extensions are supplied by the system.  The root
!   file name cannot be more than 8 characters, and spaces are not allowed
!   in path naming.
```

```
      {OutputName=D:\NALC123s\NALC123s}
!
!    EnoughSegs is a processing cutoff in terms of number of segments.
!    If the number of segments reaches or exceeds this cutoff, then the
!    program will go into its final clustering scan.  Put a minus sign on
!    this specification for a continuation run to refine an existing
!    segmentation.
      {EnoughSegs=2500}
!
!    MaxScans is an upper limit on the number of scans through the data,
!    including the final clustering scan.
      {MaxScans=12}
!
!    DO NOT ALTER THE ENSUING LINES.
!
[RUN] {DataFile} {DataBands} {UseBands} {PixelRows} {PixelCols}
[RUN] {OutputName} {EnoughSegs} {MaxScans}
```

In fact, the standard template file for the PSISCANS program not only has this annotation regarding the specifications but also additional annotation regarding the purpose of the module and the nature of the outputs. The line(s) starting with [RUN] at the end of the specification file are of fixed form and should not be altered. Curly brackets must appear before and after each specification, and there should be only one specification per line.

The PSI5 folder contains the template specification files, the executable files for the modules, and the C source code files for the modules. To use a module without the front end, one would appropriately alter a copy of the specification template file with a text editor and then activate the executable module, as for example by double clicking in the PC-WINDOWS environment. Thus, the combination of specification file and executable for each module is intended to be self-standing – even with respect to documentation. Again, it bears emphasis that specification files are to be prepared with a text editing facility rather than a word processing facility. Both NOTEPAD and WORDPAD are suitable text editing facilities in the PC-WINDOWS environment.

Glossary

Compression – Conveying aspects of image patterns electronically in a more parsimonious manner.

Context – Fragmentation, juxtaposition and interspersion of elements for patterns and images.

Contrast – Distinctiveness of elements for patterns and images relative to visual perception.

DEM – Digital elevation model.

Disposition – Distribution of position for a particular kind of pattern element over the image lattice.

Echelons – Hierarchical components of a surface relative to topological trends.

EMR – Electromagnetic radiation.

Enhancement – Facilitating the visual perception of particular image patterns.

GIS – Geographic information system for storing, analyzing and presenting spatial information.

NDVI – Normalized difference vegetation index.

Ordered overtones – Imaging a map of ranks for magnitudes of signal vectors.

Pattern picture – Portraying a mapping of patterns in pseudo-color mode using pattern properties to formulate the color palette.

Pattern potential – A measure of dispersion for pattern elements relative to other patterns.

Pattern profile – Frequency distribution for patterns in a block of pixels.

Pattern progression – A sequence of patterns produced by a progressive patterning process.

Physiography – Terrain and topography.

Pixel – A (cellular) element of a rectangular image lattice (raster) on which a signal vector is determined.

Pixel pattern – A (sub)set of pixels from an image lattice sharing a common signal vector.

Polar potential – A joint measure of diversity and abundance for a particular set of signal vectors.

Polypattern – A hierarchy of patterns within patterns.

Principal (pattern) properties – Principal components of pattern properties.

Prominence of pattern – Relative occurrence of a pattern as proportion of pixels after null pixels have been excluded.

Property pattern – A (sub)set of signal vectors from among those occurring in an image lattice.

Property points – Signal vectors plotted as points in signal space having bands as axes.

Proxy – a signal vector that is ascribed to all elements of a pattern.

PSIMAPP – Progressively Segmented Image Modeling As Poly-Patterns software system.

Randscape – A hypothetical landscape image in which all property patterns have random spatial arrangement.

Raster – Row/column rectangular lattice of cellular elements.

Remote sensing – Acquisition of signal vectors for images using energy that propagates through an intervening physical medium or space.

RGB – Red, Green, Blue.

RHII – Regional habitat importance index.

Segmentation – Partitioning an image into (sub)sets of pixels, with a set of pixels comprising a segment not necessarily being contiguous.

Signal bands – A band is one of the measured components in a multivariate signal vector.

Signature – A characterization of a pattern for purposes of algorithmic pattern recognition and classification.

Species richness – Number of species in the area of interest.

Stretch – Computationally or optically altering the range of values for a signal band.

Synoptic – Data that ascribes signal vectors to all locations within the geographic extent of interest.

Supervised classification – Classification by specifying the characteristics of signal vectors to be included in each class or category.

Training set – A (sub)set of data used to specify the characteristics of signal vectors to be included in a class or category.

Unsupervised classification – Using a general segregation strategy to determine sets of image elements, which are then assigned class labels retrospectively.

Index